What the Experts Are Saying About *Future War*

"*Future War* is an exceptionally interesting book by the foremost authority on the subject. . . . It is the best delineation of fact concerning the history and challenge of this fascinating subject available."

—Lt. Gen. Richard G. Trefry, U.S. Army (retired), former military assistant to President Bush

"The ideas in this book are a necessary first step in the way we think about alternatives for the current kinetic solutions to today and tomorrow's security concerns."

—General John J. Sheehan, USMC (retired), former commander in chief U.S. Atlantic Command, former supreme Allied commander, Atlantic

"John Alexander has written an important work on non-lethal weapons that is must reading for all those interested in the nature of future conflicts. This book is a great means to understand the technology, employment, and future potential of non-lethal weapons."

—General Anthony Zinni, USMC, commander in chief U.S. Central Command

"Adroitly blending fictional action scenarios with real-world geopolitics, *Future War* captures our imaginations and sweeps us into an innovative new kind of future conflict."

—David Morrell, author of *First Blood*

"There are big conceptual and policy gaps between the very lethal nuclear weapons that were so much a part of the Cold War and the less-than-lethal weapons that the Defense Department is exploring for the post-Cold War world. No one has bridged these gaps better than John Alexander with his book, *Future War*."

—Professor Harvey Sapolsky, director of Defense and Arms Control Studies Program, Massachusetts Institute of Technology

"This book provides a provocative, thought-producing consideration of a means of warfare that surely will become more and more prevalent in the immediate future. It behooves every professional military scholar and practitioner to study this fascinating and erudite presentation."

—Lieutenant General Trefry, author of *Parameters: the Journal of the Army War College*

"*Future War* engages the reader in considering the very broad spectrum of new missions our nation's servicemen and women are called upon to execute."

—Admiral David E. Jeremiah, U.S. Navy (retired), former vice chairman of the Joint Chiefs of Staff

"*Future War* will become the bible for those seeking answers to the role of non-lethal weapons in modern warfare."

—General E. C. Meyer, U.S. Army (retired), former Chief of Staff

"*Future War,* by Dr. John Alexander, makes an excellent contribution to this debate by addressing the numerous complex issues surrounding the development and use of non-lethal weapons."

—Vice Admiral Norman Ray, U.S. Navy (retired), NATO's assistant secretary general for defense support

"*Future War* is a major piece of work, readable as science fiction, but rigorous and informed; it is a must to understand the background and to keep up with the fast-evolving concepts and technologies in the field of NLW."

—Rear Admiral Guillermo Leira, NATO director of armaments planning, programs, and policy

"His study of the ways of taking out belligerents without the intent of killing them...should grab and impress thriller readers, techie fandom in general, and serious students of military affairs.... Coverage of this subject in relatively few pages makes the book a model of conciseness and completeness, and it is well organized, to boot. . . . Highly recommended."

—*Booklist*

"Alexander repeatedly will surprise even the best-informed reader with details of how new weaponry has already changed the outcome of military operations. . . . An impressive range of information on the newest military technologies. Bedside reading for the general dying to spend next year's budget."

—*Kirkus Reviews*

"Alexander is one of the acknowledged specialists in this field, and this is the first book on the topic written for lay readers."

—*Library Journal*

"The book offers a clear, nontechnical account of non-lethal, antipersonnel, and antimatériel weapons development."

—*Civilization*

"[Alexander] has produced an important book—even for those who think that bullets and explosives will constitute the essence of military power for the foreseeable future."

—*Foreign Affairs*

"A groundbreaking primer for the cutting-edge, less-than-lethal weapons that are already appearing on battlefields at home and abroad. Part weapons and tactics exposé and part techno-thriller, *Future War* reveals how non-lethal weapons—far more than just rubber bullets and tear gas—are pivotal in dealing with today's ethically and strategically complex conflicts."

—*African Sun Times*

FUTURE WAR

FUTURE WAR

Non-Lethal Weapons in Twenty-First-Century Warfare

JOHN B. ALEXANDER
Colonel, U.S. Army (retired)

THOMAS DUNNE BOOKS
ST. MARTIN'S GRIFFIN
NEW YORK

THOMAS DUNNE BOOKS.
An imprint of St. Martin's Press.

www.stmartins.com

Library of Congress Cataloging-in-Publication Data

Alexander, John B.
Future war : non-lethal weapons in twenty-first-century warfare / John Alexander.
p. cm.
Includes bibliographical references and index.
ISBN 0-312-19416-1 (hc)
ISBN 0-312-26739-8 (pbk)
1. Non-lethal weapons. 2. War—Forecasting.
3. Law enforcement—Forecasting.
I. Title.
U795.A43 1999
355.8—dc21 98-19466
CIP

10 9 8 7 6 5 4 3

CONTENTS

FOREWORD

THE NEW WORLD'S ORDERS

Tom Clancy

The good news is that the bad news is our fault. The United States of America has brought much to the world. The world in which we live is very different from that in which our grandfathers came to manhood. Back then it was expected that some of the young men would flock to recruiting posts and don their country's uniform, and be sent away to risk their lives on a battlefield chosen by their government, there to be shot at by men in different-colored uniforms selected and sent to the same place from another direction, and with a different mission. Some would die. Some would come home, there to be honored as the victorious guardians of something or other.

It's been that way for all men for a very long time, probably as far back as civilization goes, but it's always been a little bit different with Americans. America, remember, is the child of the Enlightenment, of Edmund Burke and Adam Smith, and the ideas they propounded. And America was a country in which even these soldiers had the right to vote, to make their wishes known to the government that selected them—and these American citizen-soldiers expected that their opinions and wishes would be heeded by someone. In other lands, the soldiers were something akin to peasants, taken from their assigned place upon the land, armed, trained, and dispatched on orders from above—as, indeed, they were expected to live their civilian lives, knowing their place and staying in it, little different to the givers of their orders from the horses they rode. The technical term, after the invention of gunpowder was cannon-fodder, the people upon whom the big guns fed. They fought because they were told. Americans, from the Revolution onwards, fought in the main because they had a personal stake in the outcome, and they fought for ideas which they mainly shared, whether it was independence from England, or the westward expansion, or to fight a war to end all war, or to stop German and Japanese aggression, or most recently, to rescue Kuwait. There has always been an ethos of American arms, commemorated in the song of the 11th Armored Cavalry Regiment, the proud Blackhorse Cavalry:

We are the Blackhorse troopers,
The finest in the land.
We fight for right
And use our might
To *free our fellow man.*

Liberation has always been the American justification for sending our sons in harm's way, to share the gift of freedom with others, the gift given us by our fathers in Washington's Continental Army. Men are more likely to risk their lives for ideas than anything else, most particularly for ideas that concern the life of the soul and give the excuse of nobility to their cause. Such ideas have often enough been perverted by others—and perhaps even by ourselves—but to send a man in the way of physical danger you need something bigger than that man himself, and only an idea fills that particular order.

He must also have a sufficiently good reason to take a life, however, and that comes hard for Americans, trained from youth to respect the lives of others as they ask others to respect their own. The United States military has since 1865 been a *firepower* army, one that uses objects or machines to take the place of human lives, because American lives are not thing to be spent lightly. A French general in World War I is said to have proclaimed of an objective, "That hill is worth ten thousand men!" (To which a battalion commander attending the mission brief is said to have remarked, soto voce, "Generous bastard, isn't he!") We have rarely, if ever, thought quite that way. Our soldiers are *citizens,* with the right to vote—they are the stuff of our government, which ultimately gives the orders, and therefore not mere fodder for cannon. So from early on, Americans have sought ways to minimize our own casualties by using technology in the place of flesh. This technique worked—it succeeded on the field of arms, giving victory to these brash colonials, and both through that example, and from the ideas the victorious Americans brought with them, so spread the idea that human life has more value than that of a horse, and so the armies of the world have learned in fits and starts to pay greater heed to the value of the expendable poker chips.

The next spillover was inevitable. If human life is not to be lost lightly, then so it is not to be *taken* lightly, because every enemy is a potential friend to be liberated—have we not brought democracy to Germany and Japan? Indeed, have we not brought democracy to Russia without the need for the global war in whose dread shadow all of the Babyboomers grew up? And so, even the lives of our enemies have value. A result of that was the American development of precision-guided weapons, smart bombs that hit their point targets both efficiently and with minimal collateral damage, the legalistic term for killing people unfortunate enough to live close to something we don't like.

The idea that human—even enemy—life has value is one that our society proclaims through its news media, ever ready to have a camera and a commentator close to the site where life has been taken, to ask questions, and show the bodies of those who manifestly are no longer a threat to anyone or anything, and then ask why those lives had to be taken.

And so, it is America's fault that the weapons that our forces wield cause results from which we would prefer to turn away. It is as true of our domestic lives. American police forces are not allowed to assault criminals, whose basic rights are established in our federal constitution, and further embellished by the evolving standards of our society. As John Alexander points out in this book, while what happened to Rodney King might have been quite acceptable in a Hollywood cop film, on morning television news it is quite another thing. The idea that even a criminal has rights which may not be violated is yet another American gift to the world, and it exists, simply, because every so often police make mistakes, and when *that* happens, it might be *us* out there, and, no, we don't want that to happen, do we?

Counter to these concepts is the harsh reality that society must sometimes employ force. Some criminals must be arrested, sometimes by force. Some nations must be chastised, sometimes by force. Just as policemen carry firearms, so must our military use weapons of great power. The need for that will not soon vanish, but as society evolves, there will frequently be a need for some middle ground, when we wish to stop people from doing things we find wrong, but to do so without creating widows and orphans in the process.

And again, America seeks answers in technology. If there is a means short of killing to prevent unlawful acts, it will strike any American as something to be pursued. If there is a way to disarm or neutralize an armed force short of slaughter, then it serves the purpose both of assuaging the American conscience and leaving behind a legacy of mercy in the minds of our erstwhile enemies, the better to serve the purpose of postwar peace and conciliation—and, by the way, the better to show the vastness of American power.

John Alexander is a former colonel in the United States Army's Special Forces, a man who knows what things are like out at the sharp end of the lance. Upon leaving the Army, he got a doctor's degree and ended up working at Los Alamos National Laboratory, thus bridging the gap between academia and practical reality. In this book he examines the future and the past that brought us to the threshold of that future. The future is a place where we will all live—consciences and all. If America is to continue her leadership in ideas, then the ideas you will find here are of value, as they will help define how America pursues her international policy objectives, how we employ force to persuade those whose objectives are counter to our own to alter their methods. Will these ideas replace the use of deadly force? No, they will not. But it's always a good thing to have more than one card to play in any game, and John Alexander is busy thinking about how we can create some new cards.

ACKNOWLEDGMENTS

As with any work in a complex area, there are many people who have contributed to my understanding of the topic of non-lethal weapons and helped provide a platform from which to speak. They range from senior generals, admirals, and policy makers to bench scientists. In developing the field and writing this book, I owe thanks to Admiral David Jeremiah, General E. C. "Shy" Meyer, General Lloyd "Fig" Newton, General Glenn Otis, Admiral Bill Owens, General John Sheehan, General Carl Stiner, General Anthony Zinni, Lieutenant General Bud Forster, Lieutenant General Larry Skibbie, Lieutenant General Martin Steele, Lieutenant General Gordon Sumner, Vice Admiral Ron Thunman, Lieutenant General Dick Trefry, Major General Cecil Powell, Malcolm Wiener, Les Gelb, Edward Teller, George Singley, Alvin Toffler, Chips Stewart, Earl Rubright, David Boyd, Bob Tolle, Ray Downs, Kit Green, John McMahon, Harvey Sapolsky, Dean Judd, John Browne, Allen Murashige, John Petersen, Colonel John Warden, Colonel Jamie Gough, Colonel John Barry, Captain Dave Carroll, Robert Kupperman, Ben Rich, Don Henry, Charles Swett, Hildi Libby and her staff at ARDEC, Ed Scannell, Steve Small, Tom Starke, Oke Shannon, Jerry Perrizo, Andy Andrews, Marty Piltch, Johndale Solem, Lieutenant Sid Heal, Master Sergeant Bud Schiff, Scott Kinkead, Pat Unkefer, Craig Taylor, Paul Kozemchak, Major John Klaaren, Major Ron Mitchell, Captain Stan Coerr, Jamie Cuadros, Bob Reinovsky, Keith Gardner, Susan Hudson, Nelson Jackson, Joe Hylan, Ron Pandolfi, Ted Handle, Marc Alexander, Steve Scott, Dennis Richburg, Jim Richardson, Jim Danneskiold, Clay Easterly, Arnis Mangolds, John Dering, Ainslie Young, John Meier, all the members of the NATO-AGARD study teams for AAS-40 and AAS-43, Colonel Andy Mazzara and his entire staff at the Joint Non-Lethal Weapons Directorate, and the others who have supported the non-lethal weapons effort.

To my friend Tom Clancy I am especially indebted for the foreword to this book, our discussions, his encouragement, and for including non-lethal weapons—where appropriate—in his books. Through his novels, the concepts have reached far more people than by way of any of the numerous technical briefings we have given.

In addition, I want to thank Daniel Tomcheff for correcting my early manuscript, my editors at St. Martin's Press, Tom Burke and Pete Wolverton, as well as my publisher, Tom Dunne. I also thank my agent and longtime friend Ralph Blum for his efforts.

To my wife, Victoria, goes special recognition, for she has patiently endured the necessary conferences and separations, plus my frustrations in drafting this work.

INTRODUCTION

General John J. Sheehan
U.S. Marine Corps (Retired)

John Alexander has asked a very fundamental policy question about the use of force. This question, while applicable for military institutions, is equally relevant for local and state governments. Consider that while soldiers in national armies have shrunk by about 20 percent, private "security" firms have grown to the point that they outnumber most national armies. We in the United States spend about $50 billion annually on private security. That amount exceeds the defense budget of every NATO nation save the United States. It is estimated that, worldwide, there are about five million military-style weapons available to "private citizens," and almost half of them are here in the United States. Whether you are a U.S. soldier deployed to Haiti or Bosnia or an urban police officer, you currently have few options when faced with a threatening situation. Rightly, we in the United States are still bound by a tradition of proportional use of force, but our adversaries are not. The ideas in this book are a necessary first step in the way we think about alternatives for the current kinetic solutions to today's and tomorrow's concerns.

PROLOGUE

WHAT'S IN A NAME?

This means that killing is not the important thing.
—*Li Quan (Tang Dynasty)*

No cars had passed the area for more than an hour. Checking his watch and noting that it was nearly 3:30 A.M., *Yuri Krasnikov slowly emerged from his camouflaged position. To ensure that no one accidentally intruded upon him, he had stayed hidden, lying motionless for nearly eight hours. His years of training with the KGB served him well. Krasnikov had been a lieutenant when the Communist government collapsed. Recognizing that his career potential had suddenly evaporated, he searched for other employment. It was the rapidly developing criminal organization, rising to fill the inevitable power vacuum, that caught his attention. It was not his first choice, but at least he would live relatively well. In addition to skills in espionage, he was accomplished in English and had traveled abroad extensively. Combined, these attributes made him a valuable asset.*

With old Soviet night-vision equipment he meticulously scanned the area, looking for any signs of police. Convinced there was no surveillance of this location, Krasnikov approached the open aquifer that flowed into the water supply for Los Angeles and surrounding southern California, opened his backpack, and extracted a sealed vial that contained a deadly biological warfare agent.

The device was cleverly designed. The sealed container would remain almost submerged but would be carried along until it entered the city's reservoirs. There, at a prescribed time, the container would disintegrate, releasing small, water-soluble packets, each containing the deadly virus. Normally, viruses would not survive in water without a buffer. Therefore, each small packet contained hundreds of microspheres encapsulated in an artificial membrane of micelles and lipids, analogous to lipofectin. These microspheres would protect the virus from water, but would dissolve when ingested and brought into contact with the stomach acids. The deadly BW agent would be released directly into each victim. Should the large container be screened by a filter, the packages would still be released and float to their destination. The intended result was to afflict thousands of unsuspecting citizens. Some would die. Many others would become terribly ill. But all in the area would lose confidence in the ability of their government to protect them.

The action was ordered in response to arrests of several members of the Russian Mafia two months earlier. They had been caught attempting to sell weapons-grade fissile material for delivery to terrorist organizations. Reportedly, the Cali Cartel was now interested in expanding their business interests in both legitimate and new illegitimate areas. With the phenomenal profits amassed from drugs, the cartel had been able to buy state-of-the-art weapons and communications equipment. They had even established their own privately funded think tanks. Their studies concluded that if terrorists posed a more daunting threat to the United States, then they could relieve the increasing pressure on the drug trade. Therefore, becoming a supplier to terrorist groups made sense. The trade would make money, and at the same time decrease the amount of drugs intercepted by law enforcement agents, thereby adding to their profit all around. To accomplish this, an uneasy alliance was established between the Russian Mafia and the Cali Cartel. The Russians would sell the critical material to the cartel, who would in turn make it available to terrorists desiring to increase their panoply of capabilities. After all, one didn't need a nuclear weapon to cause extensive damage with such substances. The mere presence of radioactive material in any blast would make area remediation extremely difficult.

Unfortunately for them, intelligence operatives had discovered the plan and established a covert counterproliferation operation. Dubbed Dark Angel, *the mission was to prevent the transfer of the nuclear material. Secret agents of the American CIA and FBI had posed as members of the Cali Cartel and offered to buy some of the nuclear material that had been stolen from the less-secure storages of the former Soviet Union. In early interactions, the Russian agents had proven conclusively that the material they offered to sell was the real thing. This revelation had moved Dark Angel to the front burner. The president had signed an executive order that authorized broad powers in removing this threat to national security. Use of force was explicitly approved. Termination of agents directly involved in the illicit trade was sanctioned. However, collateral casualties were not acceptable—especially within the boundaries of the United States.*

The undercover intelligence agents had arranged the transfer meeting on the outskirts of Fort-de-France, the capital of the Caribbean island of Martinique. Martinique, a mountainous volcanic member of the West Indies archipelago, was insisted upon by the Russians. Administered by the French overseas department, they believed it would be difficult for any foreign agents to operate there without encountering the arcane diplomatic procedures followed by the French government. The tourism industry that flourished in the islands would facilitate the multinational meeting without causing suspicion.

In a temporary storage facility at a designated time, the trade was to have occurred. Brad Banta, an Army major detailed to the Interagency Counterproliferation Task Force, was in charge of the assault team that would effect the apprehension of the Russian Mafia agents. He was ably assisted by FBI Hostage Rescue Team–trained snipers and CIA field operatives. The team had trained together extensively, and at this level, interagency rivalry was negated. Participants had demonstrated their individual skills and come to respect one another. Banta was chosen to lead this operation based on his years of Special Forces experience, which included some of the most dangerous behind-the-lines missions imaginable. He was known for innovation in very difficult situations and had been involved in covert raids against the Russian Mafia in the Ukraine. In this operations security was paramount, and neither the French nor the local government of Martinique had been informed about the impending strike. There would be no host-nation support. In fact, it was highly probable that diplomatic repercussions would follow. Due to opposition to the death penalty, the French had even recently refused to extradite convicted murderer Ira Einhorn, who had been captured living near Bordeaux. In this operation it was likely that some people would die on their island. They had also been resistant to tightening the noose around Saddam Hussein when the United States wanted to demonstrate resolve. That was in November of 1997, when inspectors were refused admittance to suspected BW weapons sites in Iraq. In the current case, some additional foreign aid directly to Martinique would soothe the protocol wounds, provided no innocent civilians were injured or killed.

At about 11:30 P.M. the meeting commenced. Once it was ascertained that the nuclear material was on hand, a signal activated Banta's response team. Disorienting flash-bang devices detonated with a deafening roar and an accompanying dazzling brilliance that caught the Mafia criminals off guard. A few instinctively raised weapons. They died almost instantly before they could begin to fire accurately. Snipers with silenced weapons had been placed strategically to eliminate any attempted resistance. Their frangible bullets were designed to enter a body and break into pieces, thus preventing high-power slugs from exiting the building and striking unwitting civilians. Those criminals who did not reach for their weapons each noticed two or three red dots illuminating their chests. Fear and common sense told them that the slightest threatening move would also end their lives.

A Russian driver stationed outside heard the commotion and attempted to flee in the van that was transporting the material. This action had been anticipated by the snatch team. A non-lethal snare capable of stopping cars traveling at rates of 60 mph had been set in place. Remotely triggered, the nearly invisible, high-strength polymer bonds intercepted the accelerating van. Within 200 feet the driver found himself unable to move forward, his doors restrained and the ubiquitous red laser designators dotting his face and head. He was a mercenary, not a zealot on a crusade. Choosing not to die today, he placed his hands outside the vehicle in full view, demonstrating that no further resistance would be attempted.

Within 30 minutes, the eight surviving criminals, bound, hooded, and sedated, were on a C-130 Hercules headed for the U.S. mainland. Temporarily, they were accompanied by the bodies of four dead comrades. Somewhere over the dark expanse of the Caribbean, long known for keeping submerged secrets, the Russians heard the rushing sound of wind as the tailgate was lowered. They were unaware of the wrapped and weighted cadavers that plummeted downward, striking the water and slipping into unmarked, watery graves. This was probably improper conduct, but disposal at sea was easier than explaining how they came to be corpses on foreign soil. The denial was plausible. As if to assuage his soul, one of the agents on board said a brief prayer for the recently departed.

Upon arrival at Hurlbert Field, home of the U.S. Air Force 22nd Special Operations Command, the prisoners were roused from their drug-induced rest, greeted, and formally arrested. They remained at a remote site in the palmetto-encrusted west Florida swamps that surround Hurlbert for several days. After extensive interrogation, they were sent on to Washington, there to be indicted on charges of terrorism.

In Moscow, the upper echelon of the Russian Mafia was furious. It was the first time so many of their colleagues had been incarcerated in the United States. Further, it meant that it would be unsafe to conduct operations anywhere in the world without concern for U.S. intervention. In actuality, they knew their extensive criminal activities had reached a point that had been determined to be a threat to national interests. Making a speech in Moscow in November of 1997, Louis Freeh, the director of the FBI, had stopped just short of making that proclamation. Continued operations had crossed the line, and the United States had demonstrated they were serious about their counterterror and counterproliferation stance.

In a bold measure, the hierarchy of the Russian Mafia decided to take a confrontational path and attempt intimidation. After all, even Israeli prime minister Benjamin Netanyahu had agreed to exchange prisoners after Mossad agents botched an assassination attempt in Amman, Jordan. The plan included kidnapping Americans working in Russia and a threat to a major metropolitan area on the mainland. When the U.S. government failed to negotiate release of the prisoners, the water supply for Los Angeles had been selected as the first target.

Kneeling beside the concrete aquifer, Krasnikov deftly started to activate the timing mechanism that would cause the BW agent to be released. Then, without warning, he heard the sound of muffled explosions followed by a second noise, that of the deploying nets that sprang forth from the command-activated VOLCANO mines. Unbeknownst to Krasnikov, agents had tailed him to this site when he was rehearsing his plan of attack. This, and his alternate sites, had all been seeded with these non-lethal derivatives of antipersonnel mines and each site placed under constant surveillance.

Before he could react, the nets enhanced with a very sticky adhesive dropped over him. Running along the polymer fiber were thin electrical wires. These were attached to a simple nine-volt battery, and a charge was automatically injected into Krasnikov, instantly interrupting his neurological motor functioning. Totally disabled, he could offer no resistance, nor could he run. Banta ran toward Krasnikov with a special foam projector. Taking aim at the BW canister, Banta directed the fast-drying epoxy stream and quickly encased the container. Even if a dead-man switch had been designed into the system, the lethal BW agent would be trapped before it could spread.

In a few minutes Krasnikov, a bit dazed and bewildered, was fully recovering in the custody of FBI agents. The BW canister, solidly entombed in epoxy, was deposited in the waiting HAZMAT truck. The response team quickly gathered all of the nets and recovered the mine casings. The following morning, prearranged grounds crews would make the area look as if the operation had never occurred. Even though Krasnikov had been captured, there was no guarantee that a second team wasn't available for another attempt. The New York World Trade Center and Oklahoma City bombings had raised awareness of the potential for terrorism on U.S. soil. For now, there was no need to scare the American public further.

Fiction? Possibly. These actions, or similar ones, will occur. The nature of conflict has changed. The emerging threats are real. The non-lethal weapons are real. The complex issues are real. Let us now explore the fascinating, and dangerous, world of Future War—in real terms.

The thunder of the future portends a unique confluence of circumstances that will distinctly and forever change the nature of conflict. Meeting these complex challenges requires—no, *demands*—the development and use of new weapons systems, ones that dramatically limit the amount of damage they cause. Traditional lethal weapons will still be necessary to defeat stubborn conventional enemies. However, new weapons, such as those described in the preceding vignette, must be brought into the standard inventory.

In the United States the official term for these weapons systems is *non-lethal.* This was formally announced in March 1996 at the Non-Lethal Defense II Conference I chaired in McLean, Virginia. Ambassador H. Allen Holmes, Assistant Secretary of Defense (Special Operations and Low Intensity Conflict), addressed the complexity of the term and provided the definition, which is as follows:

> *Non-Lethal Weapons.* Weapons that are explicitly designed and primarily employed so as to incapacitate personnel or matériel, while minimizing fatalities, permanent injury to personnel, and undesired damage to property and the environment.
>
> 1. Unlike conventional lethal weapons that destroy their targets principally through blast, penetration, and fragmentation, non-lethal weapons employ means other than gross physical destruction to prevent the target from functioning.
> 2. Non-lethal weapons are intended to have one, or both, of the following characteristics:
>   a. They have relatively reversible effects on personnel or matériel.
>   b. They affect objects differently within their area of influence.

Non-lethal, while far from being a perfect word, does provide an adequate context from which to address the issues related to diminishing the number of collateral fatalities, a factor that will take on increased importance in future conflicts. In this book we shall discuss the problematic semantics endemic in attempting to describe this emerging field. The downside of our choice of non-lethal as a descriptor is that detractors will continue to claim that no system can be designed that is absolutely fatality free—a point readily acknowledged by the definition and everyone working in the area. To some people, this term connotes that no one will ever be killed by the use of any non-lethal weapons system. Unfortunately, there are no perfect systems that can ensure that loss of life will never occur. Even marshmallows, properly placed, can kill. Rather, non-lethal weapons are designed with the intent of limiting physical damage. Nothing will prohibit misapplication or eliminate accidents, which are training and management issues.

Now, three factors have converged to make the development of non-lethal weapons essential. They are:

(1) the dramatic and undulating reorganization of the geopolitical landscape;
(2) advances in technology, especially precision guidance, that allow refined non-lethal weapons to be effective;
(3) commanders with field experience in peace support operations to establish hard operational requirements for weapons systems development.

It was not until all three factors fell into place that proceeding with a concerted effort was deemed necessary. In fact, the whole notion of non-lethal weapons is controversial and will probably remain so for some time to come. However, in many forms of future conflict, traditional military force will have severe limitations. Conventional thinkers cannot be allowed to design a military that is only capable of defeating consuetudinary adversaries against whom bombs, tanks, and missiles are the only weapons necessary. Rather, it is imperative that we envision and forge a versatile force, one capable of deterring or vanquishing any enemy, no matter how strong, yet one that is able to impose our will in complex, often illusive situations that demand restraint. Non-lethal weapons must be a part of the inventory.

Law enforcement agencies continue to use the term *less-than-lethal* when talking about weapons that incapacitate. While many note the merging of missions, currently law enforcement focus is somewhat different from the military, and their mandate for use of force is much more restrictive. They can use deadly force only to protect life. However, there are many circumstances under which physical force must be employed to take charge of a situation; protecting property, for example, or controlling a person in custody. Police are painfully aware that their actions will be carefully scrutinized whenever they use force. Therefore, tolerance for unintended fatalities approaches zero. The National Institute of Justice (NIJ) describes its less-than-lethal research and development initiatives in very broad terms, covering the entire life cycle of crime from commission of a felony through conviction and long-term incarceration. Thus, the technologies developed under NIJ range from smart weapons and protective equipment to systems that can be employed in dispersing prison riots.

Non-lethal weapons are not a panacea. In conjunction with lethal weapons, they provide military commanders with additional options but do not restrict the old ones. The same is true for law enforcement agencies. This book is not about a peace movement or handcuffing the police. It is about the difficulties and realities of force application in future conflicts and for use on the streets of America. What it means is that military forces and law enforcement agencies, both tasked with maintaining peace, are valiantly attempting to do their jobs while limiting the use of force necessary to accomplish those missions. These missions point to the urgent need for more and better non-lethal weapons alternatives.

Part I

THE RATIONALE

It was late September 1996, and the autumn sun warmed the gently rolling hills of the Italian countryside as about 150 representatives of NATO countries met to discuss non-lethal weapons. San Piero a Grado, an Italian Navy research and development center located just north of Pisa, hosted the meeting—the first of its kind—and I had been invited to open the session on technology. My message was simple: The technology necessary for non-lethal weapons either exists or could be developed in a relatively short time. The real issue was for military officers to set requirements and state the capabilities they needed. NATO soldiers who had been engaged in peace-support operations in Bosnia and other countries were now providing the real-world context for those requirements.

To understand non-lethal weapons, it is essential to understand the situations our militaries and law enforcement agencies will face in the future. If you believe that wars and crime will be the same as they have been in the past, then there is little need for non-lethal weapons. The very nature of conflict is changing. Soldiers and law enforcement officials will be challenged by circumstances almost unimaginable two decades ago. To accomplish their missions, they will need a wide variety of options for use of force. Non-lethal weapons must be part of their arsenal.

Therefore, Part I of this book describes in some detail how the future is likely to evolve. It provides the foundation for contemplating the force options that will be necessary and the rationale for the development and deployment of non-lethal weapons. Some non-lethal weapons exist. More are needed—now. Here's why.

1

WHY ARE WE DOING THIS?

Nothing is harder than armed struggle.
—*Sun Tzu (Chou Dynasty)*

Since the fall of the Berlin Wall, the notion of "non-lethal warfare" has been hotly debated in military circles. The shift from bipolar confrontation, in which *national survival* was the driving force, to a geopolitically complex world requiring regional stability and engendering transient, pragmatic relationships requires us to rethink the whole notion of national security.

In the bygone era of the Cold War, Western military forces were structured to fight the most dreaded of all battles: war in Central Europe. Our national security policy, articulated in National Security Directive NSC-68, was containment and deterrence. It was oriented solely on countering the Soviets. That thinking was dominated by the strategic triad of nuclear weapons systems: long-range, stealthy bombers; precision-targeted, underground-based intercontinental ballistic missiles; and highly survivable submarines armed with sea-launched ballistic missiles. These forces were to hold the Soviet empire in check and ensure our survival through a policy known as Mutually Assured Destruction (MAD). MAD meant that if a nuclear war erupted, we would destroy each other, and probably the world. To work, it depended on rational adversaries who would not risk total destruction of civilization as we know it.

However, plans had been drafted on both sides that could allow a war of massive forces without crossing the *nuclear threshold.* Therefore, in addition to these strategic forces, the United States and its NATO allies developed and maintained large conventional military forces comprised of modern ships, airplanes, tanks, and artillery. The driving factor was to have sufficient forces to meet a Russian and Warsaw Pact invasion head-on, stop their advance, and be able to restore the boundaries to their pre-war state. Since we always assumed the Soviets were developing new and better fighting systems, our development efforts were designed to meet threats twenty years down the road.

The Soviets were known to rely heavily on armored forces; thus, their tanks were of keen interest to us. In the early 1980s studies were conducted and reports published stating that there was a significant "armor gap" between the West and the Soviets. Urgent action was required, and a major armor/antiarmor initiative was undertaken. While we knew their new T-82 was being fielded but didn't know its characteristics, we hypothesized about the next generations, dubbed Future Soviet Tank (FST) I and II, respectively. Shortly after I retired from the army in 1988, there was even talk about FST-III. It was the perceived high-tech, future Soviet threat that was used to justify the development of most new systems. That was the focus and raison d'être of all military research and development.

We structured our military forces assuming that if we could defeat the Soviets in Central Europe, any other military engagement could be considered a lesser-included case. That means our forces would be able to defeat any other adversary. Of course, the American experience in Vietnam unequivocally demonstrated the problems with that thinking. Heavy forces could not fight well in jungles and were seasonally restricted in the rice paddies. Air and sea power provided extensive firepower but couldn't occupy and hold the territory. Politically (but not militarily) defeated, we abandoned Vietnam. The psychological impact of our experience in Southeast Asia would shape the military leadership for decades to come. Most of all, the military leadership wanted troops to be employed only when clear military objectives could be established, and the support of both the people and elected political officials was firmly behind them.

During the drawdown and restructuring that followed, high-technology weapons continued to be developed. Their worth was undeniably proven in 1991 during Desert Storm with the rapid and devastating defeat of the Iraqi armored and air forces. In the 1967 Six Day War, Israeli troops demonstrated that "what could be seen could be hit." Now we have confirmed "What could be hit could be killed," and "We owned the night." With our advanced sensor systems, including satellites, reconnaissance aircraft, and ground-based systems, we could find any exposed target, strike it, and kill it, *day or night.* Precision-guided munitions decimated tanks and other armored vehicles at a phenomenal rate.

In the eight years Iraq fought ground battles with Iran, troops learned they were safe if they stayed in tanks. In the first two to three days of Desert Storm, they learned they were safe only if they did not stay in the armored vehicles. UN air supremacy was established within a few hours of the onset of the operation. Their ground-based air defenses obliterated, the Iraqi Air Force learned instantly not to challenge the far superior UN fighters. Similarly, the state-of-the-art bunkers they possessed for their aircraft were quickly destroyed by our pinpoint penetrating munitions. The survivors were a few brave souls who dashed to the relative safety of their old archenemy, Iran.

The recent development of military non-lethal concepts arose from very lethal roots. While law enforcement has always been charged with using the minimum force necessary to restrain assailants, the post-Vietnam military embraced the concepts of overmatching enemy weapons and the use of overwhelming force. "Overmatching" meant that, system for system, we could shoot farther and more accurately than any adversary. Overwhelming force indicated that we did not believe in our age-old principle of "fighting fair"—at least not when it came to war.

When I went to Los Alamos National Laboratory in August of 1988, the Soviets were still our predominant adversary. Although we were working on weapons technologies that would temporarily disable any enemy system, the rationale was that it would be easier to disable that system than to kill it. The Pentagon had established an Office of Strategic Competitiveness under the Office of the Secretary of Defense. Headed by Dan Goure, a political-military affairs expert and son of a famed Sovietologist, the office was determined to learn how we could defeat the Soviet military asymmetrically. That meant employing methods in which we did not have to go head-to-head with their formidable armored forces.

While military leaders always felt we would eventually win any war—including Central Europe—they knew our casualty rates would be substantial. They also had to consider the problems associated with bringing our force to bear against an enemy that planned to advance 50–100 kilometers each day. The Soviet forces planned to establish and maintain a ferocious tempo of attack. The fast-moving, heavily armored forces would be very difficult to stop. The United States concluded that we would "trade space for time." That meant we would conduct a retrograde action while moving our troops to their prepositioned equipment. Then we would counterattack and restore the boundaries. Our German allies did not like the idea of trading space. They pointed out that it was their "space" we were willing to trade.

Defending Central Europe presented tremendous logistical problems. The Soviets, we believed, would come in such great numbers that even superior technology would have problems killing the multitude of tanks that would be thrown at our front lines. Their doctrine called for wave after wave of armored forces, supported by an incredible amount of artillery, to keep up constant, unrelenting pressure. Everyone on the NATO side knew it would be a tough fight—one none of them wanted to engage in.

It was in this environment that I first started developing concepts for non-lethal weapons. The initial ideas were focused on an established concept: breaking threat tempo. That meant if we could delay the Soviet reinforcements as deeply in their territory as possible, it would provide time for NATO forces to get troops and ammunition forward in time to destroy the forward-deployed Soviet military. The concept was in sync with the newly developed concept of Follow-on-Forces Attack, or FOFA, as it was known. The only difference was that instead of destroying the enemy forces, the weapons I proposed would temporarily delay their arrival at the front. Short delays deep in enemy territory would result in a cascading effect, thus producing major influence on the battlefield. Somewhat ironically, rather than being "non-lethal," the ultimate desired effect of these weapons was to increase the kill ratio in the forward battle area.

An additional issue was the population density of Europe. Cities large and small cover much of the landscape. From World War II engagements, we knew and understood the problems of fighting in cities. Going house to house, it is a very nasty battle. Since the United States was going to move light infantry into the theater, the only place it could be expected to survive a brutal armored onslaught was in difficult terrain or in cities. As I studied the problem, I was surprised to find a document written nearly twenty years earlier by Joseph Coates of the Institute for Defense Analysis (IDA). It was titled *Nonlethal and Nondestructive Combat in Cities Overseas* and addressed many of the same ideas and technologies that were beginning to form the heart of the current non-lethal weapons concepts.[1] A problem at the time Coates wrote his thesis was that the technologies he proposed were not sufficiently mature to offer practical weapons applications.

Over the next few years, the United States and its allies engaged in operations that would prove instrumental in the evolution of non-lethal weapons and concepts. We had already gone after Manuel Noriega in Panama with Operation Just Cause. Then we were involved in several operations-other-than-war, including Somalia, Haiti, Cuba, and, most recently, Bosnia. These operations provided commanders with the experience necessary to begin to formulate requirements for new non-lethal systems.

For many, the very words *non-lethal weapons* represent an oxymoron. Is not the destruction of your adversary the objective of war? While controversial, the answer to that question is, "Not really." In our fictionalized versions of war, we tend to concentrate on the total destruction of any group of people on whom we have bestowed the title of *enemy.* This was epitomized in the 1996 hit movie *Independence Day.* In the film, the U.S. President asked the alien invaders if accommodations could be made. "What do you want us to do?" asked the president. The movie alien response was "Die!" This exchange serves to show how deeply ingrained in popular consciousness the concept of physical destruction of an enemy has become.

The real objective of war, however, is the imposition of *will.* Getting an adversary to do as you dictate, not their physical demise, is the desired outcome. The Prussian military theorist Karl von Clausewitz called war "an extension of diplomacy by other means. Imposition of will, not physical destruction, is the appropriate measure of success."[2] In fact, in the long run, physical destruction and unwarranted fatalities may be counterproductive to the goals of most advanced civilizations. There are two distinct issues that come into play in termination of conflict. One is the rebuilding that is required upon the end of hostilities. Recent history has shown that it is usually the victor who bears the heavier financial burden. The domestic economies that made the transition to support a war effort must be restructured to meet peacetime requirements. Additionally, the victors often feel compelled to assist the vanquished in reestablishing internal economic and political stability. The results of such efforts are best seen in the post–World War II support rendered to both Germany and Japan. From abject devastation, in a few short decades, they have emerged as major world powers.

The second issue relates to holding grudges. Throughout the world, there are many societies that have both large families and long memories. While physically destroying an enemy in battle may preclude future attacks in the short term, it does not prevent later retaliation. Today the world is replete with conflicts based on old animosities, events that transpired decades, even centuries ago.

The events of recent years in the Balkans and the southern states of the former Soviet Union demonstrate that emotions can be repressed by force for long periods of time. There can even be a semblance of integration and civility among the factions involved. Yet once the physical repression is relaxed, the conflicts resume. Even if a majority of the personally aggrieved parties have died off, new generations, imbued by the stories of past atrocities, seem fully prepared to take up the cause as their own.

It does not take large numbers of casualties to generate enduring hatred. Consider an ambush that occurred in the Middle East in which the prophet Ali ibn Abi Talib and about two dozen followers were murdered. That incident happened thirteen centuries ago in A.D. 661, and led to the split between the Sunni and Shiite Moslems. The two sects have been engaged in conflicts, both philosophical and physical, ever since.

The violent nature of most conflicts inevitably leads to violations of established military protocols and ethical constraints. The more intimately involved in combat the individual becomes, either physically or emotionally, the more likely it is for atrocities to occur. On the ground, hostilities can easily get up close and very personal. All military forces, no matter how professional they consider themselves to be, have committed unauthorized acts of extreme violence. U.S. forces in Vietnam were no exception. Most Americans were shocked to learn about the involvement of U.S. Army troops in the My Lai Massacre. In that incident, young American soldiers, mostly draftees, willfully executed 347 unarmed men, women, and children. Herded into the ditch by the infantry platoon commanded by Lieutenant William Calley, the villagers were machine-gunned to death at point-blank range. The reason? Some of the soldiers' buddies had recently been killed in fighting in that area. Professional soldiers or not, combat evokes passion. Passion can get out of control, and frequently does. Retribution seems justified. The cycle continues.

Civilians around the world are being taught that retribution is acceptable, and sometimes desirable. It is a prevalent theme in novels, movies, and television. Justified violence has become a hallmark in Hollywood films, and is employed to gain the emotional support of the audience for acts that the hero is about to commit, albeit reluctantly. To make sure acts appear more acceptable and less self-serving, bad things happen to family members or close friends of the hero. The hero, his or her family killed or threatened, then sets about "righting wrongs" or protecting himself or herself and others. Any amount of force is acceptable, usually the more the better, from an audience perspective. Superstars such as Clint Eastwood (*Dirty Harry*), Mel Gibson (*Lethal Weapon* and *Ransom*), Wesley Snipes (*Passenger 57*), Steven Seagal (*Under Siege* and *Above the Law*), Arnold Schwarzenegger (*True Lies*), and Sylvester Stallone (*Rambo*) become the personification of revenge. They are rewarded by committed movie audiences that demand further involvement in retribution. The problem is that, while these actors always appear to solve one problem by the end of the movie, in real life other problems continue to emerge.

Passionate responses are not limited to military operations. In recent years, law enforcement agencies have come to recognize the physical and emotional responses prevalent when they are engaged in chasing fleeing felons. The situation is so common that it has become known as High-Speed Police Chase Syndrome. While meeting the exigent, and sometimes life-threatening, requirements of the situation, the police officer experiences an adrenaline rush. Once the suspect is apprehended, the officer must exercise extreme restraint, even though his or her physical system is still in a fight-or-flight mode. Failure to control the emotional stimuli can lead to civic tragedies, such as the Rodney King case in Los Angeles. Having been a deputy sheriff in Dade County, Florida, and personally involved in high-speed police chases, I can attest that this altered state is easily attained.

Contrary to current popular belief, war has always represented the controlled application of force. Over two thousand years ago, Sun Tzu, the venerable Chinese military strategist, addressed the importance of fundamental non-lethal concepts. In *The Art of War* he wrote, "The general rule for use of the military is that it is better to keep a nation intact than to destroy it. It is better to keep an army intact than to destroy it, it is better to keep a division intact than to destroy it, it is better to keep a battalion intact than to destroy it, it is better to keep a unit intact than to destroy it."[3]

Physical force is threatened, or applied, when differences cannot be resolved by other diplomatic means. The biggest problem in the application of force is that it initiates or continues the "cycle of violence" and rarely leads to long-term solutions. While differences will continue to occur and armed forces will actively participate in resolving them, options that limit violence will have inherent advantages over those that accentuate it.

Semantics have been extremely important in the discussions about non-lethal weapons and concepts. Many variations have been proposed. When I wrote my first article on this topic in 1989, I was concerned about our primary threat: the possibility of war with the Soviet Union and Warsaw Pact in Central Europe. Soviet doctrine demanded rapid advances on a predetermined time schedule, or tempo. To describe the new concepts for disrupting that enemy tempo, I adopted the words *antimatériel technology.*[4] The basic idea was to concentrate on stopping the machines of war, not necessarily on killing enemy soldiers. As the weapons applications expanded and the geopolitical realities and military missions changed, the term *antimatériel* clearly did not encompass the totality of concepts involved.

A term that experienced some favor was *soft kill.* This term inferred that attacks that limited destructive effects to soft and vulnerable nodes, while not causing catastrophic physical damage, would be made against a weapons system. In principle, it would be easier to create mild damage at critical junctures than total physical destruction. For instance, you could target sensitive electronics and cause them to malfunction without blowing up the heavily fortified command bunker that houses the equipment. Likewise, degrading optics can inhibit mobility of armored vehicles, and jamming communications can prevent coordinated assaults. The inherent advantage of the soft kill was that it took less energy to incapacitate an enemy system than to destroy it physically.

However, to many the term *soft kill* was a non sequitur. After all, things die hard, not soft. However, the term was used in several popular articles and occasionally can be found in use today. It was this term, coupled with societal references by President George Bush, that led people to talk about kinder, gentler war. Several articles touted such titles as *Killing Them Softly* and *Bang—You're Alive.* Clearly, the authors did not understand the basic concepts they were writing about.

Another derivative term is *mission kill,* which means that an enemy system has been rendered ineffective because it can no longer accomplish a specific function. In the military, there are three main functions necessary for a weapons system to be effective and accomplish assigned missions: it must move, shoot, and communicate. The Defense Advanced Research Projects Agency (DARPA) commissioned a study chaired by General Glenn K. Otis, a retired Army officer with extensive armor experience.[5] The panel recommended that mission kill be defined in the following words: "A military operating system, person, or unit suffers a mission kill when an adversary's action causes it to be unable to perform its assigned function at a time or place required." Examples of a mission kill could include a tank not being able to move, a cannon not firing, and radios not communicating. In every case, the system, while not destroyed, would be rendered ineffective in combat. Thus, a mission kill can be a cost-effective method of stopping an enemy with high-technology weapon systems.

The term *less lethal* also has been used to describe these new weapons. Less lethal implies that some level of lethal action will occur, but collateral fatalities can be minimized. It does not, however, have the same limiting connotation as non-lethal, which means that no one would be killed.

In response to concerns that some non-lethal weapons might cause injury but not kill, other terms emerged. Harvey Sapolsky, Director of Arms Control Studies Program at MIT, hosted one of the first meetings on non-lethal weapons. In noting the plethora of emerging terms, he even suggested that some weapons would be considered "*worse than lethal.*" This was derived from concerns about weapons that might blind or cause other abhorrent permanent injuries. In general these are emotional rather than factual arguments. Later we will discuss the physical effects of various non-lethal weapons, some of which do cause discomfort or short-term pain.

Despite the controversy about these weapons, senior military officials have set a framework in place. The directive assigning the title non-lethal also designated the U.S. Marine Corps as the Executive Agent for Department of Defense Non-Lethal Weapons Programs. Under the leadership of Marine Commandant, General Krulak, and the Commander-in-Chief, U.S. Atlantic Command, General John Sheehan, also a Marine, the Corps had actively pursued this assignment. U.S. experience in peace support operations has made non-lethal weapons development a high-priority issue. The Army chose to take a supporting role and let the Marines take over the program lead.[6]

Certainly, the preferred term, *non-lethal,* has serious drawbacks. Therefore, it must be understood that when force is used, fatalities are possible. When force is employed at a national level, some fatalities are likely to occur.

One approach to defining non-lethal weapons has been to quantify the number of fatalities that can be accepted or tolerated. Every study in which I have been involved has invariably made an attempt at such quantification. Some suggest that 1 percent fatalities might be an acceptable rate, while others would place the acceptable figure somewhat lower. The process is based on a well-established method for determining weapons effects. For many decades, precise measurements have been worked out for predicting the probability of casualties when rounds carrying high-explosive warheads are used. Weapons experts calculate the Probability of Kill (P_k) for each weapon, based on explosive power, fragmentation, distance from detonation, and the amount of protection afforded people at the target site. Such a formula can also project the probability of nonfatal casualties and even divide those figures into serious and minor injuries. Having such an explicit methodology has been useful in making critical decisions about weapons development and employment.

Unfortunately, the models are not very useful when it comes to questions concerning non-lethal weapons. The weapons involved do not lend themselves to such simple equations. Also, no one is ever prepared to provide a firm number that constitutes the "minimum acceptable number of casualties."

Another very similar definition was put forth by NATO Advisory Group on Aerospace Research and Development (AGARD) studies involving non-lethal weapons. The definition was, "Non-lethal weapons are those weapons that are designed to function in a manner that degrades the capabilities of matériel or personnel and yet avoid unintentional casualties." The study went on to state, "With such systems, there should be few, if any, fatalities or serious permanent physical damage to humans. It is acknowledged that some casualties may occur due to accident or misuse of such systems. It is recommended that non-lethal weapons systems always be supported with lethal capabilities."[7]

There has been far too much discussion of semantics, especially by lawyers, academics, and political observers. The consternation generated has allowed those who are not supportive of the basic concepts to obfuscate the real issues. It is important that the fundamental issues and concepts are understood in the context of current geopolitical realities. In many future military missions, as well as with police protecting our citizenry, use of deadly force will necessarily be minimized. The name applied to that task is not really important. Providing appropriate weapons options for field commanders and law enforcement officers is.

2

ARE WE THE WORLD'S POLICE FORCE?

I see a whole lot of Albanias, Haitis, and Mogadishus in the future.
—General John J. Sheehan, USMC (Retired), Former Commander in Chief U.S. Atlantic Command

Somalia: Restore Hope

The fighting intensified as night came on 3 October 1993. The 75th Rangers were attempting to exfiltrate from an abortive mission to capture the local warlord, Mohammed Farah Aidid. Since that afternoon when they were pinned down, the Rangers had been judicious in their shooting, attempting to limit civilian casualties. They were, after all, in a densely populated section of Mogadishu, which was known as the most dangerous city in the world. Now there were a number of wounded Rangers trapped with their dead comrades only a few hundred feet from the encroaching enemy. Helicopters had attempted a rescue earlier in the day. One was shot down only a few blocks away. In that action, there would be two Congressional Medals of Honor awarded posthumously. These would go to Master Sergeant Gary I. Gordon and Sergeant First Class Randall D. Shughart, members of the elite Delta Force who had risked their lives, had endured withering fire, and had eventually been shot down almost within sight of those whom they tried valiantly to help. That day the effect television can have on policy was graphically demonstrated. For all Americans to see, there would be the repulsive images of a dead, partially stripped American soldier being dragged on ropes through the streets of Mogadishu like a slaughtered animal. By night's end, eighteen Rangers were dead, surrounded by more than 300 dead or dying Somali citizens, civilians and guerrillas alike. The Rangers and Special Forces, equipped with some of the latest in weapons technology, had been trapped and slaughtered by what was regarded as a ragtag militia.[1]

The operation had been hastily launched based on poor intelligence. Bill Cullen, senior CIA official, later admitted that they had relied on a single source of questionable veracity who went by the code name Lincoln. In military and intelligence circles, Aidid had been dubbed "Elvis" and Lincoln was reporting another "Elvis sighting." The incentive was intriguing; Elvis and twenty of his senior lieutenants were to meet mid-afternoon in a high-rise hotel. The U.S. military commanders could not wait for the cumbersome intelligence process that required routing back through CIA Headquarters in Langley, Virginia. The Rangers and Delta Force soldiers lifted off on what was expected to be a ninety-minute raid. What they didn't know was that Italian forces, stationed at the other end of the flight strip, had made a secret deal with Aidid. As soon as the helicopters took off Aidid was warned of the attack giving him plenty of time to set a deadly trap.

It would not be until years later that the truth about the enemy would become known to the public. Some of these supposed "militia" members were in fact seasoned veterans who had successfully fought another technologically superior force: the Soviet Army in Afghanistan. They had been recruited and equipped by Osama Bin Laden, an extremely wealthy dissident from Saudi Arabia who had supported jihad in several Islamic countries. From his perspective, the United States falsely stated our humanitarian purpose, and we were, in fact, attempting to subjugate Somalia.[2]

The seasoned Somali combat veterans had been closely scrutinizing American tactics. They knew they could not match forces in direct combat so they intently searched for vulnerabilities that prevented the U.S. forces from massing their overwhelming firepower. It was in our technical superiority that they found the Achilles Heel they were looking for—helicopters. While helicopters offered great mobility, observation, and firing platforms, they could be hit by ground fire. Therefore, Aidid placed his experienced fighters, armed with rockets and machineguns, on the roofs of nearby buildings. Since he also knew who was informing on him, Aidid found it easy to place the information about the meeting and lead the Americans into the trap. During this operation the Somali guerrillas shot down two U.S. helicopters, each in an untenable location, and changed the course of the conflict.

There proved to be many more casualties of that misbegotten operation. Field commanders had asked for, and had been denied, additional firepower. Specifically, Secretary of Defense Les Aspen had personally turned down the request to provide a small but effective armored unit on the ground in Somalia. It was a decision that would cost him his job, only shortly before his unexpected fatal heart attack. Additionally, there was debate about sending AC-130H gunships that could have provided a ring of fire around the trapped troops. In the end, the gunships were not deployed. Those decisions cost the lives of Rangers, who had been caught without adequate reinforcements.[3,4]

Restore Hope was termed a peace support operation. It had begun ten months before with the quirky noninvasion during which U.S. Marines stormed across the sandy beaches of Somalia shortly before dawn on 9 December 1992. There they were greeted by the glaring lights of the news media cameras who had been tipped as to the time and location of the landing. The Marine amphibious force had been left out of Desert Storm, deployed at sea as a decoy so that Saddam Hussein would believe an invasion over the shores of Kuwait was imminent. Their very presence had tied up large numbers of Iraqi troops, who were convinced that the invasion must come from the sea, not across the seemingly impenetrable desert. Now their entrance into Somalia provided a high-profile operation that attracted more unfavorable attention than expected. Despite our declarations to the contrary, to African nations, and many others, it looked like an invasion, although we claimed it was only a routine method for insertion of troops. That supposition was hard for many to believe since, for the week prior to this event, the U.S. Air Force had been flying Pakistani troops into the commercial airport near the capital. Over 500 soldiers had been flown in and simply walked off the C-141 Starlifter aircraft without incident. Now the Marines had landed, taken charge, and Restore Hope was in full swing.

The mission was to disperse food and alleviate the massive starvation that had claimed tens of thousands of lives. The supplying of food was only one small part of the problem. Somalia was near a state of anarchy. The incessant feuding of fourteen local warlords had disrupted the flow of food to the starving people. Seemingly oblivious to the massive suffering of the local population, these warlords withheld vital supplies in order to concentrate power in their own hands. The warlords were supported by poorly trained, but modestly equipped, troops. Both the United States and the former Soviet Union were to blame for the availability of weapons. During the Cold War, this section of Africa was deemed to be of strategic significance. We chose sides, one supporting Somalia, the other Ethiopia. Then, strangely, during the middle of the confrontation, we changed sides. The result was that both the United States and the former Soviet Union sent large numbers of small arms to the area. The warlords, whom we alternately supported, hoarded the weapons and ammunition, then used them for their personal endeavors. By modern military standards, these were systems that could be easily overwhelmed in conventional battle. But peace support operations do not qualify as conventional battles.

As the humanitarian aspects of Restore Hope got underway, the U.S. forces, now bolstered by troops from the U.S. Army 10th Mountain Division, were confronted with both mines and sniping. These problems quickly took on a serious tone as a jeep was destroyed by a land mine placed in the road, killing the American soldiers and an accompanying civilian technician. Increasingly, the troops came under fire from snipers shooting from crowds. The snipers had the support of the local civilian populace, who became de facto "willing hostages." Having only limited weapons options, the U.S. forces countered snipers with rifles, machine guns, and occasionally helicopter gunships. Many civilians, the ones we were there to help, were killed. On 5 June 1993, soldiers loyal to Aidid attacked Pakistani peacekeepers, killing twenty-five. It was then that Ranger units, known for their very aggressive tactics, were brought in to assist.

Slowly, the mission transformed from support of humanitarian aid workers to suppression of the warlords. This gradual shift from one objective to another came to be known as "mission creep."[5] Now troops were engaged in conducting raids, a job much more compatible with the kind of units involved in the operation. Instead of focusing on protection of food distribution points and supply lines, the troops were now pursuing warlords. This was a method of preempting attacks on themselves, and it fit with operations the Rangers were trained to conduct.

The alteration in mission should not have been a surprise to anyone. There is an old saying, "When you don't know what to do, do what you know." The troops involved in Restore Hope had only minimal training in humanitarian operations. After all, in the U.S. military, very few soldiers had any extensive experience in peace support operations. Rangers are trained to function aggressively, but for short duration. They represent the spearhead of American combat forces. Their motto is "Rangers lead the way." The Rangers did what they knew how to do: conduct raids. It is a classic example of the wrong unit with the wrong equipment. The fault was not the Rangers', but of the senior officials who sent them to Somalia and then failed to provide adequate support.

The public outcry, largely stimulated by the evocative pictures of the desecrated body of Staff Sergeant Bill Cleveland being dragged through the streets, caused an abrupt shift in policy. That image was not consistent with our concept of a humanitarian operation. Quickly, U.S. forces departed, only to be replaced by other UN troops.

Later it was learned that the tragic ambush of the Rangers should have been avoided. In September, President Clinton had asked former President Jimmy Carter to intercede in Somalia. Carter knew Aidid and had established a personal relationship with him. President Carter had already secured a promise from Aidid to negotiate a settlement to the escalating conflict between him and the U.S. forces. Given the impending diplomatic solution, no one can satisfactorily explain why the military still had orders to capture him. This serious breach of common sense had strategic implications and directly cost the lives of the brave U.S. servicemen. However, we were not through with Somalia.

United Shield: Somalia Round II

By the end of 1994 more than 130 UN peacekeepers had lost their lives attempting to end the starvation in Somalia. What had begun as a humanitarian mission—providing food—had evolved into "peacekeeping" and even "nation-building." Mission creep indeed. Peacekeeping involved keeping warring factions separated, and later, nation-building efforts included a rough attempt at creating an infrastructure that could be self-supporting. That meant attempting to rehabilitate the economy, promote national reconciliation, establish a police force supported by a legal system, and other measures.

One of the trickiest operations was attempting to disarm dissident groups. Basically the mission failed, and after spending in excess of $2 billion, the international community was ready to go home. But getting out might not be as simple as just picking up one's toys and going home. There were those who wanted the UN forces to stay for their protection. This would be a friendly but determined group. There would also be elements of the militia that could threaten the troops during their withdrawal. As the forces diminished in strength, they would become more and more vulnerable. The last ones out would be in the most danger.

The United States agreed to enter the contest once again. Operation United Shield was to provide the covering force for the United Nations' evacuation forces. The 13th Marine Expeditionary Unit, Special Operations Capable, would constitute the unit on the ground. The task force was placed under command of Lieutenant General Anthony Zinni, who had been the director of operations for the first mission. For this operation, there was sufficient planning and training time. It was also determined that new, non-lethal weapons would be made available to the force. These systems were "off-the-shelf," as there was not sufficient time for development of new systems tailored for the mission.

Limited non-lethal weapons were introduced to the task force. Low-impact rounds for twelve-gauge shotguns, sticky and aqueous foams from Sandia National Laboratory, and a low-energy laser system were provided. In addition to troop training, the news media were alerted to the new weapons systems, and they were brought in and shown the troops training with the new systems. The plan was to let the world know that we were bringing alternatives to lethal weapons, ones that we would not be afraid to use. It was reiterated that if U.S. troops were fired upon, they would return fire with their conventional systems.

Operation United Shield was executed without a hitch. The Marines went ashore and covered the UN forces as they left. Although the Somalis followed relatively close behind the departing forces, they kept a safe distance. In a few instances, when they came too close, the low-energy laser was employed as a spotting system. The visible light placed on suspected snipers was sufficient to intimidate them. No shots were fired, and all of the forces, UN and U.S., departed safely. Later, Lieutenant General Zinni stated that although the non-lethal weapons available to him were limited, he would never go on another peace support operation without them. He also proclaimed an urgent need for more such weapons.[6]

There may have been another, more intangible, loss in Somalia: the loss of innocence. On 23 August 1994, I was invited to brief members of the Defense Science Board on non-lethal weapons technology. The location was the idyllic Beckman Center of the National Academy of Sciences on the campus of the University of California at Irvine. One member of the board was retired four-star general John Foss, former commander of the U.S. Army Training and Doctrine Command. After the Somalia operation, he had interviewed a large number of troops. When he asked the infantrymen if they knew whether they ever killed anyone, they invariably answered, "Yes." He went on to compare those with interviews of combat veterans from Vietnam. When asked the same question, they would usually say, "I don't know." In Vietnam, a lot of shooting was indiscriminate at targets hidden from sight in the jungle or at a distance, and the soldiers could not ascertain whether it was their shot that killed the enemy. In Somalia, when countering snipers, the fighting was at close range and our troops saw the people, mostly innocent civilians, fall. One situation was called combat, the other a humanitarian mission. Clearly, non-lethal weapons offer a chance for soldiers to avoid facing such dilemmas in the future. The psychological burden of killing innocent, or even tacitly supportive, civilians should not be thrust upon young troops when viable alternatives are available. Those already so encumbered know the price is just too high—and you never stop paying.

Uphold Democracy: Haiti

Since 1937, Haiti was controlled by dictators of one manner or another. For several decades it was François "Papa Doc" Duvalier, once self-proclaimed "President for Life," who ruled Haiti with an iron hand. This was accomplished with the support of the dreaded Tontons Macoutes, organized thugs who terrorized the local population through brutal beatings and horrific murders.[7] In 1971, nineteen-year-old Jean-Claude "Baby Doc" Duvalier succeeded his father. The reign of terror continued. With the economy in ruins, in 1986 riots broke out and finally the Duvalier regime was forced into exile.[8]

The first elections as a fledgling democracy did not go smoothly, but finally, in December 1990, Jean-Bertrand Aristide did become the popularly elected president. The limited number of the elite class remained in place. The leadership of the military had come from this class and had previously been loyal to the Duvaliers. After a brief time, the Haitian military decided that rehabilitation of the country was not moving quickly enough, so they initiated a coup. In September 1991, the military, led by General Raoul Cedras, toppled the elected government and appointed themselves as the new leaders. Although they promised to restore governance to the people, it was clear that was not likely to happen. Aristide's release from captivity was negotiated and he left the country and came to reside in the United States.

The economic situation continued to decline. Tens of thousands of Haitians attempted to flee the country, most frequently by unseaworthy craft. Their ultimate destination was the United States. The U.S. Navy and Coast Guard were placed in the waters to return people to Haiti, but the number of refugees continued to rise. President Clinton, anxious to stem the flow of indigent people, offered to establish an economic aid package in excess of $1 billion. However, the aid was conditional, based on the return of democracy to Haiti. In June, the United States, with the support of the UN, initiated an embargo preventing the importation of all but the most basic needs. Then, on 3 July

1993, all parties agreed to the conditions set forth at a meeting held in New York. However, the economy of Haiti was so poor that most of the people were barely living at a subsistence level, with an annual per capita income below $300. Thus the embargo had little effect on them. What goods there were became concentrated in the hands of the elite few who, likewise, were not being sufficiently hurt to cause them to restore a civilian government.

On 15 October 1993, General Cedras and Port-au-Prince police chief Lieutenant Colonel Michel François were due to leave their offices. Power assumed is difficult to relinquish. More murders of key opposition leaders were carried out. With patience running thin in the Clinton administration, it was decided to increase the pressure on the general and his accomplices. The U.S. Navy vessel, *Harlan County,* was sent to Port-au-Prince, only to be turned away by a mob waiting at the docks. This small but organized group of thugs handed the United States a media defeat that was transmitted around the world. As the *Harlan County* sailed away, U.S. prestige abroad was put on the line.

Finally, in September 1994, tiring of the antics of the military leaders, President Clinton decided on a more forceful approach. This time there was an ultimatum: Leave or be forced out by the military might of the United States. The Haitian leaders wavered, but didn't budge. In fact, they made claims about how their forces would defend their homeland. While they could not defeat the Americans, they knew that in the jungles they could inflict a substantial number of casualties, probably more than the American population would tolerate.

Then, an armed invasion force ready, Clinton dispatched ex-President Carter and retired chairman of the Joint Chiefs of Staff Colin Powell to meet with General Cedras and his cohorts. General Powell had many things going for him at the meetings. As a result of Desert Storm, he had an international reputation as a military strategist and statesman. As a black man facing overwhelming odds, he had risen to the top of the U.S. military. Now he was talking face-to-face with the black officers who had initiated the coup.

With the invasion force already airborne and an armada within sight of Port-au-Prince, Jimmy Carter and Colin Powell met for the last time with General Cedras. It was General Powell who detailed the military capabilities of the armada that was ready to be unleashed. This was not a bluff. The military leaders would either have to leave or die. The military officers fled the country and the American forces were allowed to enter peaceably.

However, what lay ahead would again stretch the capabilities of the American forces. They would not be fully prepared for the mission they had to undertake. Instead of fighting the ill-equipped and poorly trained Haitian military for control of the terrain, they arrived to find a country in near chaos. Even experienced combat veterans of the Vietnam conflict were not prepared for the extreme level of impoverishment they found in Haiti. Many had traveled in various Third World countries and thus were acquainted with different lifestyles and economic realities. Still, they were not mentally prepared for the conditions they found in Haiti.[9]

The mission for the U.S. soldiers was to support the reintegration of President Aristide. While Aristide had been elected by a popular majority, there were substantial numbers of people who did not wish him to return. Also, as a result of the poverty level and lack of trained police, crime was endemic. There was also great resentment among the general population for the elite class that had traditionally maintained their wealth at the expense of the people. Savage force had been used to subjugate the people. Immediately following the first revolution, there had been a wave of unbelievable violence. Members of the dreaded Tontons Macoutes had been tracked down and viciously beaten to death. Many had been chopped to pieces and then, while still alive, burned in the streets in front of jeering mobs. Haitians' propensity for using the machete as tool of choice for interpersonal conflict resolution bothered many soldiers. The situation in morgues was reportedly so bad that babies were kept in a pile. When mothers inquired about lost young ones, they were held up like dolls. The situation seemed surreal to young troops raised in modern America.[10]

Even as Uphold Democracy was beginning, retribution was taking place for the next wave of violence that had occurred after the coup. In some instances, beatings took place within view of the American forces. The rules of engagement held that they could not intervene with their lethal systems. Another media circus evolved as televised violence occurred while our fully armed troops stood on the sidelines, administratively prevented from intervening. From a military perspective, their options were very few. The troops could threaten force, but if their bluff was called, they had to back down, or, if their own lives were threatened, they could shoot. Neither option was satisfactory.

Slowly, a relative calm returned to the island nation. But the American troops were still having problems. To maintain order while a professional police force was being trained and the rudimentary elements of an infrastructure installed, the American forces stayed on the island. Since U.S. soldiers were being supplied with more food and other materials than most Haitians had ever seen, even a simple task, dumping trash, endangered our forces.[11] As odd as that sounds, it proved to be life-threatening. The starving Haitian people were willing to take high risks for access to our trash piles. A process was established in which the Haitians were kept at a safe distance while the American troops dumped their trash and retreated to their vehicles. Then a whistle was blown and the surge of scavengers was on. Surging forward, it was survival of the fittest as they fought to sift through the refuse. As concerned American officials on the ground pointed out, how can you possibly shoot someone who is merely attempting to rummage through your garbage? On CNN, it would be a political disaster. Unfortunately, the U.S. troops in Haiti had very few options available for protecting themselves without using lethal force.

Even the limited number of non-lethal systems that were introduced encountered a legal snafu. Since the weapons were not standard inventory, special authorization had to be given for them to be issued for the use by the invasion force. When that authorization was obtained, it was given to the

Marines. When they were replaced by an Army unit, the weapons could not be transferred. The Army did not have proper authority to receive or possess the non-lethal weapons.

Bosnia

With a great deal of reluctance, the United States agreed to support the UN and NATO by joining the United Nations Protection Force (UNPROFOR), entering the conflict in the former Yugoslavia. The situation in Bosnia-Herzegovina was one of the messiest in the world, but, with thousands of people being killed and the ever-growing images of savagery on television, there was a perceived need to "do something." No one in the government was sure what actions were appropriate or how to go about accomplishing such an ill-defined task. Before troops were placed on the ground, it was necessary first to have a truce or cease-fire enacted. The Dayton Peace Accords satisfied that requirement.

Unlike the operations previously discussed, the adversaries in this conflict had more sophisticated weapons, not just rifles and machine guns; both sides had artillery, tanks, and even air-defense systems. While they could not match the United States in an all-out battle, these weapons posed a very significant threat to our peace support troops in close proximity. The need for well-armed troops with lethal weapons was obvious.

However, there were many other problems for which we did not, and to this day do not, have good solutions. One prevalent issue was our protection of designated safe havens. These were geographical areas drawn up and agreed to by all factions. Unfortunately, major violations occurred frequently, and fighting took place very near these sanctuaries. It was determined that the UN forces were insufficient to protect the safe havens by military means, and that mission was not within the scope of their official mandate. It was decided the UN forces should be withdrawn. When that time came, the civilian population, the ones who would now be placed at great risk, attempted to prevent the withdrawal. They did so by simply staying physically very close to the UN forces. Of course it would be impossible to use lethal force against people whose only concern was their personal safety. Those remaining behind had every reason to be frightened. Many of them had seen the horrors of torture and murder that had been perpetrated against their friends and family members. They knew of the mass graves now being authenticated by the UN. Trusting the goodwill of their adversary was not a viable option. And yet the UN forces were not equipped to handle noncombatants who attempted to block their departure. This led to very difficult confrontations between the protected and their protectors.

Another situation without a solution involved the frequent and blatant violations of the Dayton peace accords. There were mutually agreed upon rules concerning the placement of certain heavy weapons, however, the violators would intentionally locate weapons systems close to civilians. Air-defense systems were placed immediately adjacent to hospitals because they knew the UN would not dare strike them. Tanks were positioned in civilian barns. Artillery was brought into villages. This was all to preclude our use of existing lethal weapons. Again, NATO forces were in a position of either using lethal weapons that would produce collateral casualties or ignoring the violations.

The World's Police Force

There is a great deal of debate about the appropriate role for America to play on the world's stage. Some argue that we should stay close to home. We should, they say, only use our military force when our national interests are directly threatened. The reality is that we *have been* involved in peace support operations, we *are* involved in peace support operations, and we *will be* involved in peace support operations. It is up to the president and Congress to determine when and how to apply force. These challenges will not go away. In fact, they will probably multiply with the devolution of other former nation-states into subelements. What is absolutely clear is that, to meet the challenges of the future, we urgently need non-lethal weapons options. Once developed and provided to our troops, non-lethal weapons will offer a wider range of responses to these difficult situations.

3

EMERGING THREATS

THIS DOESN'T HAPPEN HERE
—Newsweek,
1 May 1995

Shortly after the turn of the millennium, Iran made its move. After suffering years of oppression under Saddam Hussein, the Kurds in southern Iraq invited armed support from their Iranian colleagues. Somewhat beleaguered at home due to the enduring UN sanctions against Iraqi oil sales, Hussein could, at most, bluster and complain loudly about the incursion. His complaints fell on unsympathetic ears. Although Iraq had about 2,000 tanks available and moved some armored forces toward Basra, they did not heavily engage the Iranian tank columns. The Iranians, however, openly stated that they were very concerned about a possible counterattack by the Iraqi forces and reinforced their troops supporting the Kurds with several top armored divisions supported by heavy artillery. U.S. analysts voiced concern about the positioning of the Iranian forces, and diplomatic warnings were fired off. Despite assurances that Iran was only concerned about the safety of their troops that were supporting the Kurds, they continued to move additional forces into the region. With its strategy of dual containment of Iraq and Iran faltering, the United States placed troops on a higher state of alert.

That night, Iranian forces began a dash to the south. Kuwait's army and air force barely had time to get in position before they were overrun by heavy tank units. An American battalion on maneuvers in Kuwait became engaged in a valiant fight, but the onslaught of heavy tanks was too great. Although the U.S. unit destroyed more than a full Iranian Army tank regiment, their losses exceeded 230 soldiers killed and another 500 or so taken prisoner. The status and condition of those prisoners were not clear. Unwilling to make the same mistake that Saddam Hussein made in 1990, Iran immediately pushed on toward the Saudi oil fields. The military objective was Ad Dammam, the chief port along the Persian Gulf coast of Saudi Arabia. In an action similar to the blitzkrieg tactics of Germany in World War II, they quickly gained territory. So fast, in fact, that reinforcing the U.S.-supported air bases in the Saudi desert with fighters from the mainland was deemed too dangerous. Although these bases were developed and supplied after Desert Storm to support operations in the Gulf region, the attack was so swift they had to be abandoned to the Iranians. Within a week, Iranian forces had secured the entire Persian Gulf, to the borders of Qatar. The attack had been coordinated with agents inside Saudi Arabia. It also exploited the already strained relations between the Saudis and the new Amir of Qatar, who was allowed to take the state of Bahrain provided he did not support any side in the conflict. The incursion was supported by thousands of Saudi dissidents, comprised of citizens not related to the House of Saud, plus many non-Saudi nationals who had been imported for the domestic labor force.

A vast amount of the world's energy supply now lay in the hands of Iran. This situation was intolerable to the Western world. Iran's navy included three Russian Kilo class submarines and twelve patrol boats with antiship missiles. Their air-defense system included U.S.-built HAWK and Russian SA-5 and SA-6 SAM missiles, some of which had a ninety-kilometer range. Further, Iran's shore-based antiship missile defense system was formidable. Retaking the Middle East oil fields would not be a Desert Storm II. The world had learned not to allow the United States and her allies time to position forces and determine the rules of engagement. The closest friendly bases would be a long way off. An invasion would have to be launched primarily from U.S. naval assets. Even with two carrier groups, the forces were insufficient to establish a beachhead and support a large-scale ground invasion. Long-range bombers and standoff, precision-guided missiles were essential. Launching attacks from the sea and from bases in Israel, Turkey, and the newly admitted NATO country of Romania, U.S. forces began attacking almost immediately. In retaliation for base support provided to U.S. forces, Iran launched missile attacks against Israel. The civilian casualties were very high. Russia, concerned about its adventuresome neighbor to the south, provided covert support to the United States and its allies. Although sympathetic, the Russians were unwilling to provide open support for fear of exacerbating their tenuous situation with the fundamentalist Moslems within the boundaries of their former southern states.

Finally amassing adequate power, UN forces simultaneously attacked Iran and Saudi Arabia. Though a counterattack was launched against the forces in southern Saudi Arabia, UN forces were able to repulse the Iranians and quickly moved to regain the oil fields. To the rear of the Iranian forces, Iraqi forces now attacked the enemies' extended supply lines. Hussein joined in a secret agreement with the UN, in which he was guaranteed the right to sell oil in return for his support in this vital operation.

The attack against the Iranian mainland near the Strait of Hormuz received much stiffer resistance. The beachhead remained precarious for several days. Even after it was secured, long-range missiles continued to plague the troops attempting to land sufficient armored forces for the breakout operation. Despite staggering losses from the high-tech weapons of the United States and allied forces, Iran continued fighting. Even with heavy air strikes against Tehran and near-complete destruction of the central command and control system, independent armored units continued to attack Allied forces, inflicting significant casualties. As was predestined, we won! But at a very high price.

This narrative has described only one of the many possible scenarios for future conventional conflict played out in war games.[1] For the foreseeable future, there will continue to be a number of "bad actors" on the world stage. Some of them, including Iran, Iraq, and North Korea,* have large armored forces and can destabilize a significant portion of the globe. There are other, less recognized, threats to regional stability. Some suggest that China will behave respectably thanks to its expanding economy. However, China has a very large conventional military that has been constantly buying and selling advanced weapons, including submarines and other naval vessels. Submarines are not considered defensive weapons. Weapons acquisition, coupled with the interest China has had in the entire Pacific region for hundreds of years, is reason for concern.

Then there is the possibility of a reconstituted Soviet Union—or some variation thereof. Such a force would also possess large numbers of tanks and artillery. Each of these potential adversaries has placed a large portion of its GNP into military hardware. Each has an unstable economy and leadership that is hard to predict. Some have nuclear capability. All have access to various weapons of mass destruction: nuclear, chemical, or biological. Most are located in places that are not easy to reach. To meet such threats, it is essential that we develop and maintain a highly mobile, extremely lethal conventional force that can strike anywhere around the globe. It is because these conventional threats continue to require so much attention that non-lethal weapons cannot be looked at in isolation. That would be a mistake. Rather, it is absolutely necessary to understand the complete spectrum of threats and then to interject non-lethal weapons in a balanced and logical manner. The introduction of non-lethal weapons should not be at the expense of lethal systems, but as complimentary tools of war and peace, ones that are demanded by emerging situations.

In the situations described above and others like them, the projection of power and application of force would not be easy. Every potential adversary took note of how Desert Storm was conducted. They all learned that you don't allow the United States and its allies time to position their forces, es-

---

*At the time of writing, North Korea was in an extremely volatile state. Massive famine, precipitated by a combination of floods followed by drought, was taking a significant toll on the population. Kim Jong Il, known to his people as the Dear Leader, had stated that up to 70 percent of the population could die and there would still be enough people left to reconstitute the nation. Contrary to the intelligence analysts who had predicted a major political upheaval upon the death of the Great Leader, Kim Il Sung, the son, had consolidated power and seemed to be firmly in charge. With a psychological profile of Kim Jong Il that states, "The elevator doesn't go to the top floor," his actions are highly unpredictable. A high-level defector said that one of North Korea's military objectives in a war with the South would be to create at least 20,000 American casualties in order to cause us to withdraw from the conflict.[30] Another defector indicated that a limited nuclear capability was available to North Korea. It is possible that the situation on the Korean Peninsula could change dramatically between the time of writing and publication. If it does not, North Korea will remain dangerous for some period of time.

tablish massive logistic support, determine the parameters under which the conflict will take place, and then determine the kickoff time. Because those lessons were so well demonstrated, future conflicts may well call for forced-entry operations. The casualties sustained in any over-the-shore invasion—such as establishing a beachhead in Iran—would be very significant. However, if we are willing to absorb relatively high casualties, the outcome is certain: WE WIN. The advanced technology displayed in Desert Storm made that rule No. 1. That is not ethnocentrism; it is just a current, but not immutable, fact of military might.

The situations described above represent high-end conventional threats, threats we must be ready to meet. However, there are emerging a number of other adversaries, who may be far more pernicious and difficult to handle. The probability of engaging the low-end, hard-to-identify threats is far greater than a major conventional confrontation. They are areas of potential conflict for which the outcome is far less predictable. These threats come from non-traditional sectors, and some are already apparent. They include criminal and terrorist organizations that have attained so substantial a base of power that they can affect the stability of the entire area in which they operate. Through the use of terror, they frequently exert psychological influence far beyond their actual size and military strength. Thus, it must be understood that the future conventional threats will adopt many of the operational characteristics of terrorist organizations, leading to a blurring of traditional military missions and counterterrorism responses.

Today, international terrorism has become a global problem. Terrorist activities have been carried out in every continent save the Antarctic, in which there are no people living permanently. It is estimated that as many as 550 different terrorist organizations are operating around the world.[2] Today, even local public buildings have defensive mechanisms. Public parking has been moved away from adjacent lots because of our experience with car bombs. The influence of terrorism can be seen everywhere.

On the afternoon of Friday, 26 February 1993, at 12:18, terrorism announced its presence on the shores of the United States, with a massive explosion that rocked the twin towers of a 110-story building complex. Late that morning, a dark-colored rented van pulled into the B-2 level parking lot located beneath the south tower of the World Trade Center in Manhattan. Ironically, it parked near the fleet of Secret Service cars. The van contained an unsophisticated mixture of fertilizer and oil, which, correctly combined, made a deadly explosive composition. The blast produced a 60-by-100-foot crater that went three parking levels deep. It damaged the adjacent floors, knocking out all power and collapsing the path to the train station, crushing people caught there. The intense heat generated billowing smoke that rose as high as the ninety-sixth floor. With the elevators malfunctioning, people attempted to flee by going either up or down the fire escapes. Two hundred children were trapped on the upper observation deck. Office workers suffered from smoke inhalation as they struggled to free themselves.[3]

In the end, the six bodies were found and hundreds of people were reported injured. But the damage did not stop there. As an international center for financial transactions, dozens of institutions relied upon the services rendered at the site. The blast shut down the New York Mercantile Exchange and plunged a number of other financial organizations into chaos. Lower Manhattan, no stranger to congested traffic, was brought to a standstill,[4] but more importantly, our confidence was shaken.

Arrested were Sheik Omar Abdel-Rahman, a blind Muslim religious leader, and a small group of his followers. The evidence gathered showed that Rahman's followers had concocted the bomb in a leased garage and driven it to the trade center. Their capture was fortunate, as they had plans for a more extensive bombing campaign, including blocking the Holland Tunnel, which runs between New Jersey and Manhattan. Some readers may have already seen the potential consequences played out by Sylvester Stallone in the 1996 movie *Daylight,* where an explosion blocks the tunnels, which then become flooded, and this leads to horrendous traffic jams throughout the area.

THIS DOESN'T HAPPEN HERE, screamed the 1 May 1995 issue of *Newsweek.* But it did. The Alfred P. Murrah Federal Building, located in downtown Oklahoma City, was the scene of the most deadly bombing in American history.[5] Early on the morning of 19 April 1995, Timothy McVeigh parked a rented Ryder truck in front of the unprotected building. Shortly after he calmly walked away, at 9:02 A.M., a tremendous explosion totally destroyed the nine-story building. The shocking pictures became etched in the minds of all Americans. Who can forget the classic photo of fireman Chris Fields holding the badly battered, lifeless little body of Baylee Almon, the one-year-old child who had been pulled from the rubble? The death toll was officially listed at 168. Included among the casualties were thirteen children who had been attending the day-care center located near the brunt of the blast.[6]

Again, the bomb was a jerry-rigged device comprised of ammonium nitrate fertilizer, available in any garden-supply store, mixed with diesel fuel. From the extent of the crater and the severe damage to the facing wall of the building, it was estimated that the truck contained approximately 4,800 pounds of explosive.[7] The concussion was so great that it weakened the structure of buildings several blocks away.

American officials and the media immediately began making statements, albeit cautious ones, indicating that foreign terrorists may have infiltrated Oklahoma and detonated the bomb. A man with the physical features of a Middle Easterner who had been seen in the area left on a plane. With much fanfare he was intercepted in London and returned to the United States for questioning. We were sure it had to be foreigners. Then the next shock came. From the axle of the disintegrated truck a serial number was deciphered and traced to the Ryder truck rental agency. They were able to identify McVeigh as the man who had leased the truck, and soon, he and an accomplice, Terry Nichols, were arrested for their participation in the horrendous event that clearly signaled to all Americans, It *can* happen here.

Not a foreign agent, McVeigh was a homegrown terrorist, a member of the Order, part of an expanding militia movement that boasts as many as 100,000 adherents.[8] Steeped in conspiracy theories and proto-neofascist sentiments, militia groups with names like the Aryan Nations, Christian Identity, Vipers, Militia of Montana, and the Michigan Militia Corps, range from hate groups to paranoid organizations who believe that United Nations troops are about to take over the United States. Some, including McVeigh, believe the carnage at the Branch Davidian compound in Waco, Texas, was the result of a murder—murder intentionally committed by the FBI. While there are many questions about the events in Waco, a condemnation of all U.S. government employees for mass murder is clearly outrageous. And yet McVeigh stated that the bombing of the federal building in Oklahoma City was in retaliation for the Waco deaths and provides a brief glimpse into the ill-conceived motivations that may lead others to act equally irrationally.

Illegal use of explosives is far more common than most people realize. According to Treasury Department records, in 1995 there were 1,952 illegal explosive incidents in the United States. The easy availability of dynamite indicates a correlation between the amount of construction and bombings. Many of these incidents are for vandalism or revenge. Since no one is killed in the bombings, they tend to go unnoticed.

Another growing problem is the increased use of *nonideal explosives,* such as the fertilizer bombs used in both bombings and other improvised explosives. The effects vary substantially from those of traditional explosives. Ideal explosives—ones developed and commercially made for that purpose, like dynamite, C-4, and Semtec—contain well-known properties. However, the characteristics of nonideal explosives, such as these fertilizer bombs, are not yet well established, and most existing computer hydrodynamic codes are not applicable. Pharis Williams, an experienced explosives expert, is leading efforts at New Mexico Tech and working with various agencies to develop a better understanding of the effects of nonideal explosives and their influence on buildings, cars, and other items located in the vicinity of a blast.

Availability of explosives is compounded by easy access to information about the manufacture of bombs. A document titled *The Terrorist Handbook,* which contains instructions on making bombs and conducting many other nasty activities, was even posted on the Internet on 23 March 1995, shortly before this incident.

Many members of these militia groups are already concerned about non-lethal weapons being used against them. To some, these weapons are harbingers of some mysterious supranational agency that is about to be inflicted upon the docile and unsuspecting citizenry of the United States. One group, founded by an Indianapolis lawyer, Linda Thompson, has warned that helicopters under UN command are preparing to attack our cities. In her videos about the Waco confrontation, she has noted that non-lethal weapons were present and showed an unidentified device with a blinking light. Her commentary in *The Big Lie Continues* attacks me by name as a progenitor of non-lethal weapons. Those involved in conspiracy theories demonstrate, beyond a shadow of a doubt, that even many bright, and well-educated people may be swayed by specious arguments and led into unwise confrontations with duly constituted authorities. Without non-lethal options, those authorities will be left with only limited lethal responses.

The World Trade Center bombing in New York and the federal building blast in Oklahoma City attest to the devastating effects of relatively unsophisticated explosive devices. It was sheer luck that the fatalities in the trade center incident did not greatly outnumber those in Oklahoma City. Both events stunned the nation and led to changes in our daily lifestyles. You can no longer drive near the United States Capitol, park near or under many federal buildings, or enter numerous public buildings without being magnetically, and sometimes physically, searched. Intensified airport security procedures represent one of the most visible effects of these changes. Now it is not unusual to stand in line for fifteen minutes or more just to pass through the screening devices. These are clear examples of how small organizations have had substantial influence on the ordinary functions of modern societies by forcing us to take ever-increasing precautionary measures.

In January 1997, a number of letter bombs were sent through the mail to various locations inside the United States. When a few small parcels, postmarked Alexandria, Egypt, arrived at the United Nations the suspicion they aroused caused the building to be evacuated. Letter bombs do not destroy buildings but are capable of killing or maiming the unfortunate person who opens the letter, and anyone else in the immediate vicinity. This terror tactic can be very hard to trace. Consider the seventeen-year-reign of the so-called Unabomber. By mailing small but sophisticated bombs over an extended period of time and aimed at targets that seemed to have little in common, he was able to elude law enforcement officials. The Unabomber's downfall resulted not from skillful detective work but from his ego-driven penchant for writing to newspapers and taunting the police. The arrest of Theodore Kaczynski came only after his writings were published and recognized by his brother, who, after much soul-searching, turned him in. On 22 January 1998, to avoid the possibility of a death sentence, Kaczynski, a diagnosed paranoid schizophrenic, pleaded guilty to all counts against him. The bombings of Ted Kaczynski and others using these letter-size devices have caused substantial changes in how the Postal Service handles mail.

Another bombing disaster was narrowly averted in New York City with the 31 July 1997 arrest of Gazi Ibrahim Abu Mezer and Lafi Khalil, only hours before they allegedly were to attack the city's subway station at Atlantic Avenue in Brooklyn. With ties to the Middle Eastern terrorist group Hamas, intelligence agents believe they were part of a wider international conspiracy. Only a day before the scheduled incident in New York, Hamas had executed a deadly attack in a market in Jerusalem, killing fifteen people, injuring about 150 others, and derailing the peace talks between Israel and the Palestinians. Such coordinated terrorist activity signifies yet another increase in capability and violence. During the police raid on their apartment, both Mezer and Khalil were shot.[9] Had this situation devolved into a stand-off, non-lethal weapons may have been necessary to protect others in the building. By employing gas, flash-bang grenades, electrical shocking systems, or sticky agents, the police may have intervened without using firearms in a crowded building. Examples will follow later in the book.

Terrorism around the World

While the United States has only recently been confronted with large-scale terrorist events, coping with them is a way of life elsewhere. In Northern Ireland and the Middle East, terrorist events are relatively common. Egypt faces destabilization by attacks concentrated on tourists. In September 1997 gunmen opened fire on a tour bus loaded with Germans, killing ten. Then two months later came the infamous raid near Luxor in which seventy people, mostly tourists from Japan, Switzerland, and Germany, were killed and a three-hour gun battle with the assailants ensued. Cosmopolitan areas, including London and Paris, have experienced bombings by various dissident groups for decades. Elsewhere in Europe, terrorist activities have occurred in almost every country, including major events in Germany, France, Spain, and Italy.

Asia, too, has had problems with terrorism. India, with nearly a billion residents, has had planes and trains bombed. Disputes in Kashmir and Sri Lanka have led to many vicious terrorist attacks. Cambodia remains in a state near war. Pakistan is no stranger to bombings and other terrorist acts, including a bombing that killed twenty-five people on 16 January 1997.

Another Asian terrorist strike, one that had staggering impact, was the Aum Shinrikyo cult's chemical-agent attack on the Tokyo subway system at the government hub, Kasumigaseki. Selecting morning rush hour, between 8:09 and 8:13, sect members deposited five canisters containing the deadly poison sarin at key points along three major lines in the subway. The devices, disguised as lunchboxes and soft-drink containers, held simple binary chemical components that were mixed at the last moment. This allowed the perpetrators to be well clear of the scene before the noxious agent was dispersed into the crowds. The most seriously injured were left both blinded and hemorrhaging internally. Many others experienced blurred vision and nausea. Ten people were killed, more than 5,000 injured. But those are only the statistics.[10] The real damage was much greater. For those who had been trapped in the underground labyrinth, coughing and gasping as an unknown force struck them down, the residual psychological impact will remain indelibly imprinted for the rest of their lives. The widely broadcast television pictures suggested to subway commuters around the world that they, too, were vulnerable.[11] *Again, lifestyles were changed.*

The Tokyo police arrested Shoko Asahara, the "Venerable Master" and leader of the Aum Shinrikyo doomsday cult. The Japanese newspaper *Nihon Keizai Shimbum* characterized the attack as "an assault against society." In fact, it signified an attack against the world. Found in the Aum Shinrikyo weapons factory were other materials—ones used to make biological weapons.[12] Terrorism had taken a giant leap forward. The terrorist use of chemical and biological weapons, forecast by counterterror experts Robert Kupperman, Dave Smith of Los Alamos National Laboratory, and others, had come to pass.[13] In late 1997, in response to the now recognized threats, the U.S. government began training local agencies how to respond to chemical and biological attacks. The threat had become real and could no longer be hidden from the public.

In recent years, international criminal organizations have begun trafficking in a new commodity: nuclear materials. With the disintegration of the Soviet Union, we learned that their security measures were not nearly as strict as those of NATO nations. The result has been the development of smuggling activities catering to those who desire access to these critical materials. For the first time, we are faced with the problem of weapons-grade nuclear material in the hands of people not responsible to any government.[14]

In addition to terrorist organizations that are overtly bent on attacking government organizations, there are a number of other types of institutions emerging as threats. As with terrorism, crime—foreign and domestic—plays an important role in the supporting requirements for non-lethal weapons. The recent escalation of firepower employed by criminals demands a strong response. However, the venue does not allow the use of tanks and bombs. Frequently, lethal weapons will be inappropriate because of the potential for unacceptable collateral casualties. Therefore it is essential that we understand the impact of crime on the necessity for non-lethal weapons.

Organized crime has burgeoned in recent years. The numbers are staggering. In 1994, it was estimated that financial crime amounted to a one-half *trillion* dollar assault on legitimate financial systems. Global crime was called the world's fastest growing business, with profits estimated at about $1 *trillion* annually. To fill the power vacuum, some 5,700 criminal gangs have infiltrated and now dominate "every aspect of political, economic, and social life in the former Soviet republics."[15] These gangs have branched out and are believed to have counterparts in twenty-nine countries, including the United States.

Recently, Russian mobsters have begun to join forces with Colombian drug traffickers. They are able to expand the drug market in Russia while providing traffickers with advanced weapons such as submarines, helicopters, and surface-to-air missiles. Their influence was great enough for General Barry McCaffrey, the Clinton administration's drug czar, to state, "The Russians, along with the Nigerians, are the most threatening criminal organizations based in the United States."[16] Through use of former KGB agents, they have added a degree of sophistication not previously seen in organized crime.

Another major concern is the potential for Russian organized crime to make nuclear weapons available to the highest bidder. Many fear that the Russian military is disintegrating, thus making security of nuclear weapons a grave concern. People have already been caught attempting to sell nuclear material. However, even more frightening news emerged in late 1997 when

Alexander Lebed, the former Russian National Security Adviser, reported that as many as 100 portable, suitcase-resembling nuclear devices were missing. Each reportedly had a one-kiloton capacity. That is the equivalent of 1,000 tons of TNT. The American nuclear weapons community is split over the existence of such weapons. However, independent confirmation that such weapons were designed for terrorist applications was obtained from Alexei Yablokov, also a prior member of the Russian National Security Council.[17] He stated that they were made for the KGB, not the former Soviet Defense Ministry.

James Woolsey, former Director of Central Intelligence, stated that organized crime threats transcend traditional law enforcement and affect national security interests. So powerful are these groups that they can undermine the sovereignty of a state.[18] Of course, traditional tools of diplomacy are irrelevant. There are no existing mechanisms for establishing contact for purposes of negotiations. While criminal organizations are free to use any method of physical force necessary to achieve their aims, law enforcement agencies are greatly restricted. Robert Kupperman, the CSIS Organized Crime project director, said that organized crime "may place our traditional freedoms under severe test." He noted the threats were great enough to skew our financial and banking systems.[19]

When we consider drug trafficking, the problem can be put in perspective. Remember that the profits, estimated between $200 billion and $500 billion per year, are greater than the GNP of most of the 170 nations in the international system. In fact, Senator John Kerry states, "Drugs are the single best-selling product in the world today." He goes on to estimate the real net is "$1 trillion, approximately three-fifths the size of the federal budget of the United States."[20]

The difficulty in countering such threats can be understood best if you explore another example such as the drug cartel of Cali, Colombia. Though recent international efforts have significantly curtailed their activities, for many years the Cali Cartel conducted a multibillion-dollar operation that affected the national interests of several governments. Although they, like the more violent Medellin Cartel before them, have been seriously injured, the demand for a continuous drug supply will ensure that some organization will always fill the void. The problem in projecting and applying force against such an amorphous organization is formidable for several reasons.

First, the cartel is headquartered somewhere in the city of Cali, inside the sovereign nation of Colombia. Even if their exact geographic location could be determined, you cannot bomb the cartel without attacking Colombian soil and risking collateral casualties. This would be politically unacceptable. There have been years of mixed, sometimes tumultuous, relations with the Colombian government over drug exports. Still, the personal sacrifices made by Colombia's police, military, judiciary, and others cannot go unnoticed. The Colombian fatality rate in the drug wars was easily ten times greater than that of the United States, a fact lost on American radio talk-show hosts.

Launching an attack in or near Cali with weapons that might injure or kill innocent civilians would be unthinkable.

Second, the agricultural and manufacturing aspects of drug production take place in the rural and urban regions of several nearby countries. Bombing them is also impractical. Once the product is ready, it is transported via a number of circuitous routes, thus enmeshing several additional countries in the process. Virtually every nation in Central America and the Caribbean is involved. In some extended routes, drugs are transshipped through European ports. With their immense financial resources, the cartels could bribe officials in each of the countries through which shipment is made. Because of complex international relationships, often economically motivated, only a limited amount of abrasive commentary is authorized. Use of physical force is usually limited to whatever actions the host nations are prepared to take. Again, bombing by American planes is not even a remote consideration.

Third, from a U.S. perspective, the terminus of the distribution systems and the user community reside inside our borders. Many in the United States prefer to ignore this monumental aspect of the problem and blame the growers. But let's face it, it is the demand for the product that keeps them in business. Use of force against American citizens is a complicated matter. In the United States, there are absolute legal prohibitions against using federal military forces inside our borders. However, that act, Posse Comitatus, has been stretched, but not yet officially broken, in waging the so-called War on Drugs. Special Forces units have been used to watch suspected delivery points in remote areas, and have been present at drug busts. Ostensibly, they have only participated as observers. However, the presence of Navy SEALs Terry Abihai, Larry Vawter, John Gay, and Dennis Chalker in an Albuquerque Police Department drug raid in which a suspect was shot and killed brought this practice into serious question. An unnamed Navy spokesman stated, "SEALs *observe* police agencies on a routine basis." According to the suit, when Manuel Rameriz died, the SEALs were doing more than *observing.*

In the case of drug shipments, at each step in the process, application of conventional force is extremely difficult. While some raids are conducted at various points along the way, they are difficult to coordinate and have rarely been effective at stemming the flow of drugs. For law enforcement or military personnel to intercept shipments or apprehend drug dealers in environments predominantly occupied by innocent civilians requires the availability of non-lethal weapons.

This example is used only to demonstrate the complexities that will be encountered on a finite problem. The reality is that in the future, there will be more organizations with both the capability and intent of threatening the interests and values of Western states. There will be a fundamental issue that revolves around social groupings. People belong to many groups simultaneously. In addition to our family, we belong to religious, occupational, social, and political organizations. Currently, nationality, ethnicity, and religion serve as a cohesive factor. However, most people in the United States assume they are "Americans" first and foremost and that in crisis, differences are put aside. However, as unifying factors, such as national survival, dissipate, and information concerning beliefs becomes more prevalent, there will be a tendency toward association with groups having common beliefs. Such associations will be at the expense of allegiance to geographic boundaries.

Therefore, from a military and law enforcement perspective, understanding future social organization will be critical in planning for the structure, development, and application of force. Given the sensitivity and complexity of the missions assigned to defense and law enforcement organizations, nonlethal weapons will play an increasingly important role. Therefore, to comprehend those weapons requirements, it is essential that we understand the fundamental constructs of future societies.

The Nation-State

Much has been written about the continuance of the nation-state as a viable entity. In recent centuries, the nation-state has been the basic building block of foreign policy and international relations. Of course, there have always been subelements—sometimes powerful ones—such as religious groups, industries, and political organizations. Still, people have generally maintained loyalty to the country in which they reside. That meant that geography was a dominant factor in the ability to raise forces. With the advent of the Information Age, these social groupings will be defined by their belief systems than by geographical boundaries. The situation will be exacerbated by groupings that have only selected issues in common. Given access to the Internet, groups can form that have no physical connection with one another, mobilize to accomplish a task, and then disappear. The tenure of such organizations will vary based on the intensity and importance of the issue raised.

A threat could then be generated by a small core organization that enlists and manipulates others to commit acts against existing national interests. Employing dupes is hardly a new idea. What is new is the ability of a skilled leader to develop a broad base of support for a given cause, then incite those followers to action, all the while remaining anonymous. Already, many conspiracy-oriented sites exist on the Web. You will find some of the most outrageous allegations, usually without basis in fact, are repeated often enough to give the appearance of being true. Through repetition, these conspiracies take on lives of their own.

Use of federal force inside the United States is an emotional issue that will engender much debate. Any discussion must consider the balance between public safety and individual rights, will undoubtedly be ferociously debated. To be sure, criminal and terrorist organizations will seek to exploit discontinuities in our organizational structure. Through careful review of institutional responsibilities, either administrative or legal, these organizations will locate overlaps and underlaps in coverage and use them to our disadvantage.

As an example, the CIA can only operate outside American borders, and even then not against U.S. citizens. The FBI and other federal, state, and local law enforcement agencies hold internal investigative responsibilities. While there has been some improvement in cooperation, it is far from what is necessary to conduct fully integrated surveillance of criminal and terrorist groups. One need only observe the attempts at coordinating the War on Drugs. Despite the best efforts of twenty-some agencies and the expenditure of many millions of dollars, the street price of drugs has remained fairly constant, if not declined. And that is the bottom line, isn't it?

In other words, attempts at coordinated efforts to stop these organizations are far from perfect. While good intelligence is the key factor in stemming these illegal activities, the budgets of the organizations reporting to the Director of Central Intelligence keep going down. Too many in Congress are prepared to divert what was called "the peace dividend" to other budgetary areas. Believing the Evil Empire was gone, we would not need to allocate large resources to intelligence. In actuality, with the former Soviet Union we knew where to look. Now we must be vigilant everywhere, and there just aren't enough intelligence resources to go around. As we shall explore later, there is a direct link between non-lethal weapons and intelligence.

In addition to criminal and terrorist organizations, other groups may emerge that have the ability to threaten national and regional security. It will be extremely difficult to bring conventional force to bear against them. Legally constituted, international economic conglomerates wield sufficient power to destabilize some regions of the world. In some cases, their activities would be illegal if conducted in one country, but not against the law in another. Frequently it is a poor or developing country that permits these questionable activities. By parceling out operations, the conglomerate can remain marginally legal. By flexing their economic muscle, they can insure that stricter laws do not become enacted in Third World countries. Simultaneously, they engage in offshore banking, thus avoiding taxes at the national level. However, this situation cannot go on indefinitely. Sooner or later, pressure will force these conglomerates to change their practices, and at great expense to the company. At this point, it is entirely feasible that violence and other illegal activities may ensue.

Compounding the financial problem is digitization and telecommunications. Instead of closely controlled national financial regulations, increasingly, multinational companies are manipulating their fiscal resources in what has been termed a "supranational cyberspace."[22] With an estimated $83 trillion in the global financial market by the year 2000, the implications for shifting of power away from nation-states is staggering. Large sums of money can be transferred in seconds, greatly exceeding the reaction time of state regulatory agencies. Given that they are operating in a multitrillion-dollar pool, consider the implications if a large organization applies part of its financial resources against our national interests. An economic or regulatory response may be too little, too late, and traditional force would be inappropriate.

In addition to economic enterprises, there is a growing universe of service-providing organizations that are infiltrating every aspect of development in the Third World. With the exception of the International Monetary Fund and the World Bank, these nongovernmental organizations (NGOs) provide more development assistance than the entire United Nations system. With such a proportion of aid comes power. As former Assistant Secretary of State Jessica Mathews stated, "Increasingly, NGOs are able to push around even the largest governments."[23] She goes on to note that technology is providing the real clout.

Given the power of assistance providing NGOs, and what is at stake in economic development, it is not inconceivable that a hybrid will emerge. These could be NGOs sponsored by multinational firms designed to win confidence in their areas of operation. Once accomplished, the coopted NGOs would use that confidence to ensure favored treatment for the company.

We may well see unholy alliances between marginally legal economic entities and criminal organizations. In fact, the reverse process has been occurring for some time. As organized crime profits soared, they have bought their way into legitimate businesses. The practice is so common as to raise the thorny question, When does illegitimate money become legitimate?

As an example of how large, multinational companies now engage in ethically marginal activities, witness the variances in environmental and worker-safety laws. Repeated allegations have been made about American companies that move their operations to Mexico and other countries to avoid meeting OSHA requirements. Although international environmental treaties are being worked out and signed, implementation varies significantly. There are very strong forces at work that continue the exploitation of hardwoods and the rain forests. Water resources are being polluted with industrial waste by diversion through pesticide-infected agricultural land. The Golden Rule still prevails. Industries with money can get away with almost anything. Deprived societies that are attempting to improve their economic status are hard pressed to enforce laws that restrict the generation of cash. In addition to the environmental issues, Third World countries are a source of cheap labor. One of the reasons for decreased production costs is that safety standards, fairly rigidly enforced in the United States, are frequently nonexistent in these areas of the world.

In response to the need for international cooperation in solving these complex problems, and to gain multinational support for use of force, the United States and its allies have increasingly turned to the United Nations. Since 1990, there has been a dramatically increased propensity for armed intervention by the UN. In the forty-five years prior to that time, the UN authorized use of force six times. Between 1990 and 1996, it voted in favor of forceful intervention sixty-one times. The UN authorization for entry into Somalia was stated to be "on behalf of civilian populations."[24] President Bush stated that the United States entered "to protect national values." It is not the amount of violence that has changed. Rather, these statistics reflect a new attitude by the international community about the appropriateness of intervening in what previously had been determined to be internal affairs.

While considering the arguments incorporated into each of the foregoing philosophies, it seems that another principle may be underlying future social organization. The principle is that social organization will be based predominantly on belief systems, rather than geography or immutable cultural differences. It differs from social macro-organizations such as "civilizations" in that a substantially higher degree of fragmentation can be anticipated. There are likely to be frequent shifts in priority of beliefs and issues induced by pragmatism. This transience in attention to specific issues is likely to yield frequent change in interrelationships. Some social organizations will have a higher degree of permanency, but for many, internal and regional stability will constantly be threatened by emerging ideas and issues. Therefore, future application of force requires the availability of a broad range of weapons—both lethal and non-lethal. A more complete model for understanding future social structures and organizations is found in Appendix A.

Implications

If this hypothesis about fluctuating social organization is only partially correct, the implications for national security are enormous and will not be well received by the current defense and intelligence communities. Then Secretary of Defense William Perry, addressing the issue of cooperative security, stated that members should limit military force to what is necessary to defend their territory. He recognized that such agreements will not prevent all violence and that some peace support operations will be required. There are more conservative views. Those who embrace the "Colin Powell Doctrine," that the armed forces should only be used decisively and with the clear support of the American people, will find the future very troubling. The reality is that there will be a need to use force more frequently than we would like to admit, and rarely will the adversary or situation lend itself to the application of the overwhelming military might available to the United States. Rather, the use of force will be intentionally measured, and the outcome not clearly defined. Claude Inis, Jr., in reviewing future missions, notes that in addition to the defeat of malefactors, one military mission will be to limit violence and mitigate its effects.[25]

Societal structures of the future may be very different from those of the past. While nation-states will appear to be the dominant lawmaking bodies, other groupings will wield power equal to, or greater than, many countries. Since the end of the Cold War, we have observed the devolution of several nations into what are termed "failed states."[26] This breakup is frequently along traditional ties to "homelands." However, these homelands often transcend current geographic boundaries.[27] Robert Kaplan, in his intriguing article "The Coming Anarchy," addressed the global implications of overpopulation, crime, disease, and tribalism. He suggests that these concerns will place severe strains on existing governments.[28] Coupled with vastly improved communications technology, the potential for dissension is very great. I predict that social groupings of the future will be based more along belief systems than geography. People currently self-identify with multiple groups. In the future, fragmentation will increase. Being pragmatic, individuals will align with groups that support attitudes and beliefs that they see as personally beneficial.

We may also see the power that the wealth of a single individual or a small group can have on international politics. A very public example occurred on 19 September 1997 when Ted Turner, founder of Cable News Network, gathered a group of United Nations officials and other diplomats and announced that he was personally donating $1 billion to a foundation to aid refugees and children, clear minefields, and fight disease.[29] While providing this generous donation, he was in effect setting policy for the UN.

Access to the Internet and other modes of telecommunications make it possible for people to aggregate over very long distances. In fact, groupings can occur wherein the individuals don't really know one another and have never physically met. Already a number of unfortunate murders have occurred when trusting individuals attempted physical contact with a person with whom they had become acquainted via the Internet. Consider now that groups of people can be mobilized to action, again without being sure of the true intent of the organizer. Due to their nature, these groups can dissolve quite rapidly once their issue is resolved or the mission accomplished. I believe such groups will be organized and prompted in actions that will be adverse to the national interest of the United States and other countries. Our ability to apply force in such circumstances is very limited. Traditional force may not work. Intelligence procedures and force options must change. Inherently, these changes will face questions of fundamental rights of groups and individuals.

The emerging threats to national security range from state-sponsored terrorists, internationally organized criminals, transnational organizations, and fundamental religious organizations, to groups searching for homelands, groups that are economically deprived—sometimes starving masses—and people who believe they are aggrieved socially. These threats are both internal and external to current sovereign nations. Many do not have fixed addresses. While traditional armed forces are required for the conventional threats cited, we are woefully unprepared to meet these rapidly emerging situations. The use of force cannot solve root problems such as the population explosion, extreme poverty, and illness. However, if nations remain viable they will undoubtedly continue to view force as a solution to problems, one to which they will frequently turn. Therefore, it is absolutely imperative that new weapons options be developed as quickly as possible. A high priority must be new families of non-lethal weapons.

4

LAW ENFORCEMENT

Go ahead, punk. Make my day.
—*Clint Eastwood as Dirty Harry*

At about noon, 19 April 1993, flames were first noticed. The structure, a crudely constructed wooden building, quickly became a blazing inferno. With horror, a stunned nation watched on live television the immolation of eighty-two people, including twenty-five innocent children. A few Davidian followers, burned but alive, stumbled out of the building. So intense were the flames that within the first few minutes everyone knew that no one else would come out alive. Despite controversy, it cannot be denied that the event was precipitated by an FBI assault on the Branch Davidian ranch near Waco, Texas. So infamous is this event that *Waco* and the *attack on the Davidian compound* are almost synonymous in the minds of Americans.

The confrontation had begun fifty-one days previously, on 28 February, with an ill-conceived raid by Alcohol, Tobacco, and Fire Arms (ATF) agents. Instead of arresting David Koresh, the local leader, on one of his frequent local trips outside of his residence, it was decided to take him when he was surrounded by more than 100 loyal followers. At about 9:45 A.M. that Sunday, more than 100 heavily armed ATF agents, hiding in a caravan of cattle trailers, entered the grounds known to the Davidians as Mount Carmel Church Center. Without warning, they rapidly exited the vehicles and approached the main building. At first, Koresh opened the door to talk with the agents. Then shots were fired, and he ducked back inside, closing the heavy, metal door.

Directly assaulting the compound, the ATF agents were met by well-armed, and previously alerted, members of the Davidian sect. Gunfire was exchanged for more than an hour, with thousands of rounds being fired, many randomly. David Koresh and several followers were wounded, and five other church members died in the firefight. What would have been relatively minor weapons charges escalated to murder when four of the ATF agents were killed in their abortive raid. Battle lines were drawn, and an exhaustive siege began.

It has been speculated that the ATF wanted a high-profile bust to improve their declining budget appropriations. They did get a high-profile case—but one in which the reason for the very existence of the organization was brought into question. Even other law enforcement officials characterized the ATF as a bunch of "trigger-happy cowboys."

The FBI relieved the ATF of responsibility at Waco, and the siege began in earnest. Members of two of the FBI's elite Hostage Rescue Teams (HRTs) were brought in to take control of the situation. Immediately, they surrounded the compound, sealing entrance and egress. From their hidden positions, highly skilled snipers would observe the Davidians for the next fifty days. Armored vehicles were borrowed from the nearby Fort Hood Army base. These were used to crush the cars that belonged to the Davidians and for maneuvering close to the buildings.

Within a few days, electricity and phones were cut in an attempt to isolate the compound. In fact, during the first assault, Koresh had used the 911 emergency number to talk with the local police about the ongoing raid. It was through these conversations that a cease-fire was negotiated and the ATF allowed to retrieve their dead and wounded. Based on their understanding of the Book of Revelation in the Bible, the Davidians had an apocalyptic philosophy and had stocked up on the necessary supplies to be self-sustaining for many weeks. While a few church members did choose to leave Mount Carmel Center, most remained with Koresh.

In addition to attempting to increase physical discomfort for those remaining inside, psychological measures were initiated to pressure them to give up. Both light and sound were applied with the intent of inducing sleep deprivation.[1] For their part, the FBI agents brought in loudspeakers and commenced to play a variety of disquieting sounds. In addition to Tibetan chants, bad music, and jet planes, they also played the frightening noises of rabbits being slaughtered. Tanks were also used to fire percussion rounds periodically. Stadium lights were set up so that the house was illuminated around the clock.

The acoustic war did not go well for the FBI. In what became known as the "battle of the speakers," Koresh had his followers turn on their own music. Their speakers being stronger, it was Koresh who kept the surrounding Hostage Rescue Team members awake and uncomfortable. Both light and sound can be effective non-lethal weapons, as we shall see shortly. This application is not a good example of such capabilities.

A non-lethal weapon, CS gas, was employed on the fateful 19 April assault. This gas is known as a riot control agent (RCA) and is normally used for crowds that get out of hand. Most military personnel have been exposed briefly to this gas during training on the use of protective masks. A powerful lachriminator, a small whiff is usually sufficient to convince soldiers that their protective gear really works, and to protect themselves from prolonged exposure. In addition to rapid tearing, uncovered skin, especially if moist, can be irritated by the gas. In short, unless you are properly protected, CS is extremely uncomfortable to be around. In the Seven Mountains area of the Mekong Delta, where I commanded Special Forces team A-421, powdered CS was dropped on terrain to prevent use of the area by the Vietcong. In other words, the residual effects were believed to be strong enough to keep a persistent enemy away for several months. At Waco, the FBI would use the same chemical substance on children in a confined area.

It was known that the Davidians had some protective masks. It was also known that the children would be left unprotected, as the masks were too large to fit snugly on their faces. Attorney General Janet Reno, who had inherited the Waco mess when she took her oath of office, was repeatedly assured that CS was non-lethal. It appears that those providing that advice were not aware how CS would affect the people in the enclosed house. This would not be a brief, unpleasant exposure. The intent was to pump in CS gas *for hours.* Since children had never been exposed to CS for such a long period of time, the advice rendered the attorney general was, at best, speculative.[2] Based on the information provided, and in the belief that children were in danger, she approved the final raid. It was a decision Reno later regretted, one she would carry with her until the day she dies.[3]

At about 6:00 A.M., armored vehicles began punching holes in the walls of the residence and injecting CS gas directly into the building. Rounds containing CS were also shot into the building. As with much of the Waco planning, fundamental errors were made. In this case, the FBI had failed to take into account the weather. The wind was blowing at about thirty-five miles per hour. Thus, much of the CS gas blew away. In order to get the gas into the interior of the building, it was necessary for the armored vehicles actually to enter the structure with the body of the vehicle. As a result, it is believed that several people were crushed in this attempt to inject the CS.

Although the stated intent was a non-lethal intervention, a variety of factors produced lethal and very tragic consequences. The devastation went far beyond the loss of eighty-two lives. For many, trust of federal law enforcement was totally forfeited. This single act probably did more to promote the cause of militias and increase their membership than any other in the history of the country. Waco is a name that, despite congressional hearings, trials, and investigations, still rings in infamy. It stands as a clear example that even non-lethal weapons must be used intelligently. As for the financial costs in this case, survivors and relatives have entered wrongful death suits in excess of a billion dollars.

Unfortunately, Waco does not stand alone as a recent example of excessive force by law enforcement. The standoff at Ruby Ridge is yet another incident in which federal law enforcement used lethal force with tragic consequences. Although the number of participants was smaller, many of the same ingredients existed as were seen in the Waco fiasco.

The year before Waco, in August 1992, Randy Weaver, an avowed white supremacist, and his family were also involved in a standoff with federal law enforcement officials. The parallels in these cases are startling. The warrant against Weaver was relatively minor: failure to appear in court on a weapons charge. After observing the Weavers for months, overwhelming forces, including armored vehicles, were brought in. On 21 August, Sammy Weaver, a fourteen-year-old child and son of Randy Weaver, went out to retrieve his dog that was barking at the marshals who were attempting to surround the house surreptitiously. Sammy found his dog shot by the intruders. As he responded to his father's call to return home, Sammy was gunned down from behind by Marshals Degan and Cooper, who were hiding in the nearby woods. The following day, on 22 August, Vicki Weaver, unarmed and holding her ten-month-old baby to her bosom, was shot in the head and killed by Lon Horiuchi, lead man on the FBI HRT sniper team. She fell inside the cabin, half her face blown away, still clutching her child. When the truth finally came to light through investigations, *The Wall Street Journal* reported the circumstances of her death, stating that "Horiuchi assassinated" her. Kevin Harris, a friend of the family, was wounded at the same time, and later died of those wounds. The next morning, the FBI agents, not yet aware that they had killed Vicki, called out to her with taunting remarks.

Other parallels abound. False statements were used to justify the raid both before and after. No warnings were given. Taunts—immature and unprofessional—were made by the federal agents. Excessive deadly force was used. Following the death of Marshall Degan, the rules of engagement were illegally changed by Richard Rogers, leader of the Hostage Rescue Team that had responded to support the U.S. Marshal's Special Operations Group, and Duke Smith, Associate Director of the U.S. Marshal Service. It should be noted that Kevin Harris, who shot Degan, was later found by a jury to have fired in self-defense. The posthumous finding vindicated Harris and made his death a murder committed by the government agents, not a justifiable homicide in the line of duty.

The costs of the Ruby Ridge incident? Four dead, including Marshal Degan, Sammy and Vicki Weaver, and Kevin Harris. There were court costs, investigation expenses, congressional hearings, and $3.1 million paid by the U.S. Department of Justice in a civil settlement with Randy Weaver. And, intangibly, there was a tremendous loss of faith in federal law enforcement by millions of Americans.[4] So egregious was the conduct of the various agencies involved in this operation that the Justice Department's final report stated that the FBI's rules of engagement at Ruby Ridge "contravened the Constitution of the United States." They also said that the change in rules of engagement encouraged the use of deadly force.[5] In an extremely unusual turn of events, late in 1997 criminal charges were brought against FBI sharpshooter Lon Horiuchi. While state charges were filed, the case was dropped in federal court because of his position with the Bureau. Federal District Court Judge Edward Lodge ruled Horinuchi was acting within the scope of his federal authority.

Waco, Ruby Ridge, and hundreds of other similar incidents with state and local law enforcement agencies have demonstrated the results of excessive force when used in attempts to apprehend criminal suspects. The costs in lives, dollars, and public trust are enormous. As David Boyd, Director of Science and Technology for the National Institute of Justice, has pointed out many times, when it comes to force, the police officer today is not much different from the days of the Wild West. He, or she, now has a club and a gun.[6] Not much else has changed.

Law enforcement officials have been interested in non-lethal—or as they prefer, less-than-lethal—technologies for quite some time. In 1986, Attorney General Edwin Meese with the support of FBI Director William Webster convened a major conference on the topic at the FBI Academy at Quantico, Virginia. The conference was chaired by James "Chips" Stewart then Director of the National Institute of Justice. Since then, there has been constant interest in the field. Research and development for law enforcement is very different than that for the military. There is no central agency that provides requirements and money to contractors to develop the technologies needed by police. Instead, companies interested in selling police equipment must spend their own money to develop the product. Unlike the military, which buys huge quantities, the distributor must sell the item to each police force individually. Most people are not aware that there are over 7,000 separate law enforcement agencies in the United States alone. We tend only to think of the larger departments. However, more than 90 percent of those 7,000 police departments have fewer than twenty-five people on them. Their budgets are tight, and new equipment is paid for at a premium. For many departments, serious trade-offs have to be made when choosing between more officers and new equipment.

David Boyd has been instrumental in spreading the word about what law enforcement agencies need. His office has also become a quasi-clearinghouse for testing equipment and informing agencies about the capabilities of technology that is now available. He has also been a driving force in obtaining and providing what limited funding there is for the development of new technology that can be placed in the hands of officers within a relatively short time. Non-lethal technology is among the items he sponsors. The National Institute of Justice has an extremely broad definition for non-lethal and less-than-lethal. They have a "cradle-to-grave" interest that does not stop with the apprehension of a criminal suspect at the scene of a crime. Rather, they are interested in technologies that also encompass safe transportation of the prisoner to jail, both short-term and long-term confinement, and even monitoring of individuals on release programs. Of course, included in the incarceration periods are technologies to extract inmates unwilling to cooperate and to quell minor disturbances or even prison riots safely.

But it is with the individual police officer on the beat where the rubber meets the road. Law enforcement officials today are faced with a paradox. Americans love the "Dirty Harry" cop, as personified by Clint Eastwood in the movies. No amount of violence that befalls the villain is unacceptable—

in general, the more the better. After all, we feel he deserves it. But, in real life, the opposite is true. If a police officer shoots a suspect, it is frequently the officer who is put on trial. In many sectors of our society, distrust of law enforcement runs deep, very deep. Long gone are the days of a fleeing suspect felon being fair game. Now lethal force may be used if the officer's life, or that of another person, is in immediate danger. Even those criminals guilty of the most heinous crimes are expected to be brought to justice—alive. In fact, suits over alleged misuse of force are so common that most police academies today spend a majority of their time teaching, so that they can demonstrate officers were trained properly, to limit their liability. From a legal perspective, the government must be able to prove that training was provided on a host of specified subjects. This is done at the expense of necessary tactical information that will keep the officer alive while on patrol.

The proclivity for suing should not surprise anyone. We are, after all, the most litigious society in the world, and the deep-pockets theory applies to any government agency. Almost any use of force by police is likely to result in a civil suit against the department, even when the force was thoroughly justified. Worse, in some jurisdictions, the officers, acting in good faith, may be sued as individuals when the state has not enacted laws to protect them.

For the police officer making unexpected contact with an armed criminal, decision-making time is measured in fractions of a second. Most police shootouts occur with the participants standing less than twenty-one feet apart. An agile, attacking criminal can cover that distance in less than two seconds. Therefore, for street confrontations, police officers tend *not* to favor an alternative weapon to their pistol. They just do not have time to be switching equipment in the middle of the engagement. Even advocates for non-lethal systems do not believe these systems can, or should, replace the officer's primary weapon: the pistol. When it comes to determining how much force is necessary in any given instance, nothing is more important than *training, judgment,* and *experience.* Tactical decisions must be left to the patrolman. It is his or her life that is on the line.

However, many situations develop more slowly. For the angry or disturbed individual who may become confrontational or dangerous, there may be time to consider optional weapons systems, especially when backup arrives. For suspects armed with knives, clubs, or items other than firearms, non-lethal weapons are viable alternatives, provided the officer can maintain his or her distance while using the system. Currently, some departments allow officers to carry pepper spray or Mace for subduing suspects. Tasers and other electrical-shock systems are also carried by some agencies. With both these technologies, extreme care must be taken to ensure adequate training of the police officer. There is a fine line between causing submission of an unwilling suspect and inflicting punishment. We all saw the latter in the widely televised Rodney King apprehension.

Capturing fleeing criminals, a mainstay of Hollywood action movies, has changed dramatically in the past few years. In most high-density metropolitan areas, high-speed pursuit is restricted or prohibited altogether. New department regulations have come about due to the number of innocent bystanders who have been killed or injured in crashes caused either by the suspect or, in some instances, by police officers. Career criminals have learned that, in many jurisdictions, if they run when a police car attempts to pull them over, they will be safe—at least for the moment. A few technologies have been fielded to assist in stopping criminals in cars. Usually these are in the form of spikes that can impale the tires of the suspects car. A number of other inventions are being developed but have yet to be made available.

Any device designed to stop an automobile must also keep the afflicted vehicle from going out of control. This is a very difficult task when you have a 3,000-pound object moving at a high rate of speed. Placing an immovable object in front of the car is rarely done now. One such incident that occurred when I was with the Dade County Sheriff's Office demonstrates what can happen when such a technique is employed inappropriately. The Miami area has many bridges. During one chase, the police called ahead for a bridge to be raised, thus trapping the suspect. Instead of raising the bridge to an erect position, the operator raised the opposite side about three feet. Unfortunately, the fleeing subject didn't realize his dilemma until he was on the crest of the bridge. It was too late to stop, or even slow down. In effect, the bridge operator had imposed a death penalty for a traffic violation.

Consider the problems caused when excessive force was used in a barricade situation in Philadelphia on 23 March 1985. At 5:35 A.M., the Philadelphia Police Department began the attempted eviction of an undesirable group of people known as MOVE, from a row house at 6221 Osage Avenue. Extreme caution was in order. In August of 1978, in a similar situation at another location in the city, a gunfight had taken place, resulting in the death of Police Officer James Ramp and the wounding of three other officers and four firefighters. Although several of the MOVE members went to jail for murder, the rest moved to the Osage Avenue location. The members of MOVE were not welcomed by the community, since they had been constantly harassing people in the neighborhood for several years. They had extremely unsanitary habits, including leaving garbage and fecal material around the area, and they would harangue the neighborhood over loudspeakers they installed. Internally, they fortified the house and constructed a bunker, complete with gun ports, on the roof.

Armed with warrants, the police cordoned off the area. Long before dawn, utilities crews cut the gas and electricity to the house. At 5:35, Police Commissioner Gregore Sambor announced on a bullhorn that the MOVE members had fifteen minutes to surrender themselves.[7] No response came from the house. At 5:50, insertion teams, armed with explosives, entered the houses adjacent to 6221 and prepared for forcible entry. Seeing what was transpiring, the MOVE members fired on police. Although the police had come with some non-lethal weapons—gas, smoke, and water cannons—the fight immediately escalated to warlike levels. In the following ninety minutes, the police fired more than 10,000 rounds at the building. During that time, the explosive charges were detonated, breaching the inner walls. Tear gas and smoke grenades were fired into the building for hours. The battle was so intense that the police had to call for resupply of ammunition from the academy. Rarely do police shoot-outs involve resupply missions.

As the hours wore on, patience became thin, and frustration took over the minds of city officials. The mayor of Philadelphia, W. Wilson Goode, told television crews he had issued orders to "seize control of the house . . . *by any means necessary.*" (emphasis added) "Any means" became a bomb, one that was constructed near the site. A state police helicopter was brought in, and at 5:00 P.M. the mayor approved the bombing. At 5:27, the bomb was dropped on the roof but failed to destroy the bunker. A small fire started and it was decided to let it destroy the bunker before it was put out. However, things did not go as planned. The fire did not force the group into the open; they remained inside the building. Soon the roof caught fire. Next the fire spread beyond that building and was quickly out of control. One house after another became engulfed in the inferno. By 9:34 that evening, the situation caused by the fire was so severe that the sixth alarm was sounded. Before the fire was extinguished, an entire city block was burned nearly to the ground. Sixty-one houses lay in the smoldering ruins, sixty belonging to innocent families. Two hundred and fifty people were now homeless and deprived of most of their earthly belongings.

Except for two badly burned people, the MOVE members stayed inside the burning building. Of those who remained, eleven people, including five children, perished in the blaze. Twelve years later, a suit was brought against the city of Philadelphia. A settlement was reached between the city and surviving MOVE members and for their relatives who had died. The award totaled $4.4 million in damages. But the counting had just begun. The city had to rebuild all the houses it had destroyed. The cost in construction: $14.5 million. Then there was replacement of personal property, psychological counseling, overtime, and many other related costs. The bottom line cost for the city of Philadelphia was a whopping $30.4 million.[8]

Every year in the United States, there are a number of hostage/barricade situations. In these events, a criminal takes one or more hostages, is trapped in a building, and refuses to give himself or herself up. Fortunately, few are as serious and protracted as the MOVE confrontation. While in many cases the hostage is known to the criminal, there are others where innocent bystanders become victims. It is in these scenarios that police departments call upon the services of their highly trained SWAT (Special Weapons and Tactics) units and hostage negotiators.

Given the time it takes for hostage/barricade situations to develop, they make an excellent case for non-lethal alternatives. In fact, many SWAT teams already have some non-lethal weapons available. In the movies, SWAT members are usually portrayed as psychologically impaired, bloodthirsty snipers itching to put another notch on their high-powered rifle. The reality is that incidents involving SWAT teams rarely result in loss of life. Unfortunately, successful operations do not attract the same media attention as when things go bad; thus, the myth is perpetuated. The range of non-lethal weapons currently available includes low-impact bullets, foams, nets, lights, noise, and gas grenades. Each item will be discussed in detail in later chapters. While currently available, they are rudimentary devices and as we have seen, not always sufficient to curtail the loss of life. Substantial improvement can, and should, be made. Hostage/barricade situations make one of the strongest arguments for further development of non-lethal weapons, especially ones that can get entrenched people out of buildings.

Once an arrest has been made, the police must get the suspect to jail. Transporting an unruly suspect from the place of arrest to the local jail can be a dangerous proposition for the police officer. Even with steel separations between the back seat and the driver, handcuffed prisoners can still raise havoc. Drunk, high on drugs, or psychologically agitated, they can become very violent, kicking on windows and doors, or just beating their heads against the divider. One non-lethal means of restraint in police cars includes simple but effective air-bag technology. Inflation of the system slowly pushes the prisoner backwards but allows him or her to breathe. Not elegant, but effective.

Maintaining order during incarceration is a major responsibility for local municipalities. Once a prisoner is under police custody, the law enforcement agency is responsible for his or her physical safety. As the result of too many cases in which prisoners were severely injured or died while in custody, watchdog groups have been established that fight for prisoners' rights. Unfortunately, members of the citizens' groups rarely observe the outrageous behavior that prisoners sometimes demonstrate.

Especially in city and county jails, corrections officers often face criminals who are embarking on their career. They tend to show a bravado that is tempered by long-term sentences in state penitentiaries. Fights between prisoners or assaults on guards can erupt without a moment's warning. These must be quickly put down, as they can spread to adjacent areas. My son Marc served four years in Palm Beach County corrections and described how these situations are handled. First, a sufficient number of guards respond to the cell block. These tend to be officers of his size, six two or more, and weighing well over 200 pounds. Assembly takes less than two minutes. Then all prisoners in the cell block are warned, "If you don't want a piece of this, get on your bunks." When the teams go in, they go in hard and fast. There is no time for a prisoner to be considering whether or not he wants to stand against them. If you're not on a bunk, you're fair game.

Assault teams then aggressively enter the cell, usually armed with electrical shields. Spark gaps are set so that the hissing electrical arcs snap and crackle, thereby demonstrating that the power is on, and to evoke fear. If a prisoner comes in contact with the shield, the charge is sufficient to cause his legs to buckle and he slides down the wall like Jell-O. Those still resisting are slammed against the wall or the floor, cuffed, and unceremoniously removed by five-man teams. If a guard has been attacked and injured, it is often difficult for corrections officers to resist ignoring protocols, so extreme caution must be observed. For larger riots, soaplike foam technology can be very effective. These foams will be discussed later with other non-lethal weapons options for suppressing disturbances.

Unfortunately, dealing with potentially dangerous prisoners on a daily basis tends to harden corrections officers. Too frequently, they are forced to take verbal and occasionally physical abuse from prisoners and turn the other cheek. Corrections officers also meet many minor offenders who have the unfortunate experience of landing in jail for a brief period of time and will never return. They are usually scared to death and cause little fuss. However, it is the recidivistic individuals who provide problems that tend to etch the officer's psyche. When I served a year in corrections in Dade County, Florida, I felt just as confined as the prisoners. I left every day, but still felt the repression of the walls. Under constant tension and given provocative circumstances, it is very easy for officers to lose emotional control temporarily. Therefore, it is imperative that we provide non-lethal tools that minimize the potential for danger to both prisoners and corrections officers.

One last and very important situation is of constant concern to all police agencies. That is the possibility of major riots. In Chapter 3, "Emerging Threats," we also addressed this issue, but it bears repeating. We are not talking about college students drinking too much and getting out of hand. We are addressing groups who feel disenfranchised, have access to weapons of various kinds, and have neither fear of nor respect for police agencies or other forms of duly constituted authority. Many senior officials believe there is a serious potential for large-scale unrest in the United States. Other nations also face this possibility. Police need non-lethal options. More powerful handguns and machine guns, while necessary when dealing with gangs and organized crime, are not the answer in restoring civil order. It is here that light, sound, foams, smokes, irritants, and even malodorous substances may be in order. The requirements exceed the capabilities of any single department. I believe that non-lethal equipment packages should be acquired at the national level and maintained regionally. In the event of a disturbance, the riot package would be distributed to the local agencies for employment. Such a process would require that state and local officers be trained on a constant basis.

Law enforcement requirements differ from those of the military. However, for the past few years they have been cooperating on the development of a limited set of non-lethal weapons. For law enforcement, the tolerance for accidental death approaches zero, as compared to the marginal acceptance by soldiers on peace support operations. If the police are to live up to their motto, *"To Protect and To Serve,"* in an environment that is both increasingly dangerous and less forgiving, they must have new tools. Technologies are not the driving issue. It is the refinement of non-lethal and less-than-lethal weapons that provide the extremely low margin of error necessary. Once that is accomplished through a coordinated effort among the National Institute of Justice, state and local law enforcement agencies, and weapons developers, then the economic concerns will be met. As we have learned, it will be too expensive for states and municipalities *not* to have this equipment. Our endemic lawsuits will see to that.

Part II

TECHNOLOGIES

It's cats and dogs," exclaimed the president of the NATO Industrial Advisory Group, or NIAG. I had just finished briefing him on non-lethal weapons and concepts. Clearly, he was having trouble integrating the breadth of technology involved into a single concept. The problem is not unusual when people are first exposed to the concept of non-lethal warfare. As managers, we group thoughts into simplified packages for ease of understanding and storage. Unlike conventional weapons systems, with which most military people have experience, non-lethal weapons incorporate an extremely wide variety of technologies. Generically, they cover chemistry, biology, physics, electrical engineering, acoustics, and information technology.

The definition of non-lethal weapons is focused on the objective rather than the description of the system. To understand the range of capabilities provided under the non-lethal rubric, you must explore each of the areas of technology listed above. In fact, only a few non-lethal weapons were initially designed for that purpose. More have come from examination of existing technologies developed for other applications, but that can be modified for use in a non-lethal weapons role. Innovation has been dependent on evaluation of existing capabilities that are then applied in a new manner.

This section presents a relatively detailed description of the supporting technologies and system concepts necessary to understand non-lethal weapons. Based on these extant and emerging technologies, we can provide the weapons needed by our forces today, and for their future missions.

5

ELECTROMAGNETIC WEAPONS

Let there be light.
—*Genesis 1:3*

Lasers

Mogadishu again! Two years after we had tucked tail and run for political reasons, U.S. troops returned to Somalia. The new mission was precise and short. Operation United Shield was to cover the extraction of all remaining United Nations troops in Somalia. For the first time in any military operation—riot control excepted—the U.S. Marines publicly announced that they had been issued non-lethal weapons. Though rudimentary and limited in number, it was the first-ever deployment of these systems and posed the first real test in the field.[1] Many questions were raised about the introduction of these weapons. What capabilities did they add to the arsenal? How effective would they be? And finally, would the use of non-lethal weapons adversely impact our troops' safety? Among those non-lethal weapons were the first fielded tactical directed-energy (DE) systems—weapons that shoot photons, not bullets—and they will change the face of the battlefield forever.

The nights in Mogadishu were close, hot, and oppressive. Tension was palpable. The local warlords knew the foreign troops were leaving. Although they had fought continuously among themselves, the UN departure would create a power vacuum—one each wanted to fill. There was prestige to be gained. Displaying bravado in the face of the overwhelming firepower brought by U. S. forces could be later translated into power. Both the indigenous brigands and the Marines remembered their last encounter. Certainly, the Marines wanted no repetition of the ill-fated Ranger raid to capture Aidid. The following is an actual account of the real-world applications of eye-safe lasers.

Lieutenant Robert Ireland was an unlikely candidate for this mission. An Air Force laser jock from Phillips Laboratory at Kirtland Air Force Base, he was attached to the Marine mission to use some of the first battlefield lasers in a non-lethal role. During Desert Storm, two AN/VLO-7 STINGRAYs, a prototype countersensor laser system developed by Martin Marietta, had been deployed on Bradley infantry fighting vehicles. But the American corps commander, Lieutenant General Freddy Franks, refused to allow it to be used for fear of blinding Iraqi soldiers.[2] Lieutenant Ireland's lasers were different; they were eye-safe. Solid-state laser technology advanced considerably between 1991 and 1995. This doubled neodymium YAG (Nd/YAG) laser produced efficient green light at 532 nanometers wavelength. The collimated beam of coherent light could be used for several purposes, including target detection, target designation, and deterrence.

The illuminating green spot of Ireland's laser swept across the darkness, searching for signs of possible trouble. Snipers had plagued U.S. and UN forces since they first arrived in Somalia, and it was feared that they might try to move closer under cover of darkness. The laser allowed the Marines to see much better than relying on their third-generation night-vision goggles.

Suddenly, about a mile away, suspicious movement was spotted. After sweeping past the site, Ireland pointed the laser beam back to the area where activity had been noticed. There, inside a building, a small group of men was working feverishly. Within a few seconds, the observers noticed what appeared to be a mortar being installed. Ammunition was being piled nearby. The Marine squad leader quickly noted the location and radioed for a helicopter gunship. With pinpoint accuracy, a HELLFIRE missile could be put through the window and forever silence the mortar. Of course, it would probably bring down the entire building. Under the existing rules of engagement, that was acceptable.

Then, as the laser shone through the window on the busy mortar crew, a strange thing happened. The men noticed that they were being immersed in an eerie green light. One man came to the window and raised his hands, signaling surrender. Within a minute, the entire team was standing bathed in green laser light with their hands raised. There would be no need to destroy the building and risk collateral casualties. The mortar crew had been effectively deterred.

This was not a lone incident. Ireland also had a prototype red diode laser that operated at 650 to 670 nanometers wavelength. Instead of the green light, this laser put out a thin red beam, much like the laser pointers that are now popular in high-tech military and civilian briefing rooms around the world. Again, the Marines were using their night-vision equipment and scanning the night for sniper movement. Their diligence was rewarded. Several hundred meters away, they spotted a four-man team cautiously making its way through the dark streets. Only one man appeared to be armed. Quickly, the red spot was placed on the man who was carrying the sniper rifle. Surprised, he stopped in his tracks and didn't attempt to run or hide. Instead, he put down the rifle and raised his hands. Although only the rifleman was illuminated, all of the men with him also raised their hands, gesturing surrender. In the end, there were no U.S. casualties during Operation United Shield. While only a small part of the operation, the deterrent value of low-energy, eye-safe lasers was proven to the military.

Actually, the basic concept is not new. It has been seen in the movies several times in the past few years. The usual sequence is a prison riot during which an unruly leader notices a red dot on his body and suddenly becomes quite cooperative. Getting personal is an excellent deterrent! There are new aspects of laser weapons that make them attractive to the military. They have become smaller, lighter, dependable, and sufficiently sturdy to withstand rough handling in combat situations.

Currently available are several man-portable laser weapons. Their downside is their weight, since these systems weigh more than forty pounds, most of which is the battery. In order to deploy the laser, the soldier must relinquish his primary lethal weapon, an option that is not at all attractive to an infantryman. However, these lasers can be effective against sensors and visually acquire targets. In preliminary air-defense field tests, it was demonstrated that a single soldier could visually track a fighter aircraft and place the beam in a manner that prevented the pilot from attacking. At present, the benefits of a specialized countersensor laser system do not compensate sufficiently for the loss of hard firepower. The military counterparts of television phasers are a long way off.

However, laser weapons that can do damage to eyes not only exist, they are for sale on the open market. For approximately $1,350 you can buy a Class 4 (serious damage) laser pistol on the Internet. The developer claims the 500-joule pulse input, 3 joule output laser powered by 12 VDC, 1.5 amp batteries will provide up to 150 shots. They further claim that this device is a high-energy hole-burner. The Nd:Glass rod will not cool efficiently, so it will take a few seconds' time between shots. For the opponents of antipersonnel lasers, the issue of whether or not to field such systems is now mute. It was not an agency of any government that put them in the field, but the greed of corporations.

It is not just Americans who are openly marketing blinding weapons. In 1995, the Chinese displayed a laser weapon called the ZM-87 Portable Laser Disturber. The literature available described the laser as being designed to dazzle and blind up to ranges of 3,000 meters. The ZM-87 was being made available to armies around the world.[3]

The potential use of lasers as warning systems has merit. Around the world many people remember the tragic incident in which, on 3 July 1988, the U.S. Navy Aegis cruiser *Vincennes* downed Iranian Air flight 655, killing 290 innocent civilians. According to official records, Capt. Will Rogers III and the crew of the *Vincennes* believed the Airbus to be an attacking fighter, and fired two SM-2 antiaircraft missiles. Scrutiny of the case later revealed that the captain had lied about his location and other aspects of the incident.

The billion-dollar, high-tech warship, so sophisticated it was known as *Robocruiser,* was designed to operate on the high seas and defeat the most sophisticated weapons. Tensions in the Persian Gulf were high prior to the attack. Instead of being in international waters, as initially claimed, *Vincennes* was actually in Iranian territory, apparently trying to provoke a fight.[4] Immediately prior to the missile firing, the crew had been concerned with some small surface craft nearby that they believed might attack them. Small attack gunboats with outboard engines were not a usual threat for them.

Shortly thereafter, Captain Rogers was informed of a radar contact. Even though the flight characteristics were wrong, the radar operators stated it was a possible Iranian fighter aircraft. Distracted and ill informed, the decision was made to fire the missiles at the aircraft, even though positive identification had not been made. For Captain Rogers, there were no weapons alternatives between killing and not killing.[5] As the plane disassembled and large pieces rained from the sky, officers on the bridge of the USS *Montgomery* knew it was not a marauding fighter that had been hit. Soon, dozens of bloated, smashed bodies were floating in the warm Gulf waters. They were displayed graphically by news media around the world. Instantly, in the Middle East, the image of the Great Satan had real meaning. The incident may have provoked the terrorist bombing of Pan Am 103 in December 1988.

In 1994, NATO's Advisory Group on Aerospace Research and Development (AGARD) commissioned a study titled "Non-Lethal Means for Diverting or Forcing Non-Cooperative Aircraft to Land."[6] A driving factor for this study was the increasing implementation of no-fly zones and some tragic experiences in attempting to enforce them from the air. The *Vincennes* incident was one of those reviewed as part of the study. Another was the 14 April 1994 downing of two U.S. Army Black Hawk helicopters by U.S. Air Force fighters, costing the lives of twenty-six soldiers and civilians from the United States, Britain, France, Turkey, and their Kurdish bodyguards.[7]

The helicopters, flying over northern Iraq in support of UN Operation Provide Comfort, had a preapproved flight plan. Both pilots from the 159th Aviation Regiment, Captain Patrick McKenna and Chief Warrant Officer 2 Michael Hall, were experienced in flying this area. Their APX-100 Mark XII Identification, Friend-or-Foe (IFF) systems should have been squawking when they were intercepted by two F-15 Eagle fighters patrolling the no-fly zone. Though there had been many problems with this IFF system, the circling AWACS should have identified the birds as American.[8] Instead, without clear visual identification, the fighter pilots called them in as Iraqi, Soviet-made, Hind helicopters. Mistakenly, the AWACS cleared the fighters to engage. With their deadly air-to-air missiles, they ended the lives of those on board the helicopters, and the careers of the AWACS crew as well as their own.[9]

In both instances, the attacker failed to have positive identification before firing lethal munitions. There are a number of reasons for aircraft to transit established no-fly zones. Only one is illegal military operations; the others are usually economic. The task of the AGARD study—to attempt to control noncooperative aircraft—was one of the most difficult imaginable. *Star Trek* notwithstanding, we do not yet have tractor beams. However, it was determined that many things could be done.

One of the recommendations of the study was to use eye-safe lasers as warning mechanisms. A beam would be placed on the cockpit of the unidentified aircraft at an oblique angle by the interceptor. The laser would not take away the pilot's vision, but it could not be ignored. If the aircraft was not about to engage in combat, it is assumed the pilot would acknowledge the warning and respond as instructed. Failure to comply to the laser warning could provide demonstrable proof that the aircraft was hostile, thus making it subject to lethal consequences.

Battlefield lasers have applicability beyond deterrence. They can also operate in the near infrared (IR) range of about 800nm wavelength. Since many night-vision systems can see in the near IR, use of such lasers can substantially enhance our ability to see at greater distances, especially on very dark nights. In the visual spectrum, eye-safe lasers can also be used to prevent target acquisition by an enemy. The light from the laser is so bright that it is impossible for a soldier to look through the sights of a weapon to fire at it.

Lasers are among the most controversial new weapons. Later, we shall address the legal aspects of their use. However, threats to eyeballs or testicles will evoke an emotional response. Any weapon that is designed to harm either will not be treated rationally. Despite calls for banning lasers, this is not likely to happen. Their use as range-finders and as guidance for precision munitions makes them far too valuable to give up. Both of those attributes were key to our success in Desert Storm. Eye-safe laser weapons will provide additional operational capability.

Other countries have also developed, and fielded, non-eye-safe laser weapons. These weapons are readily available on the open arms market. Therefore, it is essential that our troops be provided with adequate eye protection whenever they engage a technically sophisticated enemy. The typical laser countermeasure is to employ a notch filter. This means that light at specific frequencies is blocked from coming through, but also degrades the soldier's ability to see. If a number of lines are blocked, then the soldier is practically blinded by the protective device.

So far we have limited our discussion of lasers to those that operate at a single frequency, and that are relatively easy to countermeasure. However, work has been done to create lasers that operate at more than one wavelength. There are two technical approaches that may be used. One is frequency diversity. These lasers would have multiple, discrete frequencies being transmitted simultaneously. The other approach is a tunable laser. That is, a laser that allows the frequency to be adjusted as desired. Both systems greatly complicate the enemy's ability to employ countermeasures since the frequencies can span the entire visible spectrum.

Battlefield Optical Munitions

On the horizon is a family of new battlefield optical munitions (BOM) that may change military concepts about directed-energy weapons. Instead of a heavy, dedicated DE weapon system, these munitions can transform standard infantry weapons, such as the M203 or M79 grenade launcher, into directed-energy systems. In recent years, prototypes have been developed and proven effective against optical sensors.

The 40-mm munition was first developed by Marty Piltch's group at Los Alamos National Laboratory. Each round contains a small amount of high explosive, a noble gas, reflectors, and a dye rod.[10] Unlike existing grenades that are propelled to the target and then explode, these rounds detonate in the chamber of the grenade launcher. The explosion excites the gases, which are reflected into the dye rod. The plastic dye lases and light is emitted. It is only light that leaves the barrel. However, it is so bright that it can dazzle optical sensors that are staring in the direction of the weapon. As infantrymen are vulnerable to heavy weapons that employ target acquisition and tracking optical sensors, this round provides them with a means to defeat such systems. The beam can be shaped so as to project a cone of light. This will cover a larger area, although the energy deposition will be lower per cm^2. Still, the energy is sufficient to dazzle the sensor temporarily while the soldier evades detection.

An innovation in the BOM family is the development of plastic dye rods across the visual spectrum. Any color light may be projected at very high intensity. Therefore, simply by randomly distributing rounds, frequency diversity is achieved. Countermeasures to a frequency-diverse weapon are very difficult and not simple to achieve.

An omnidirectional light source, known as an isotropic radiator, was also developed and tested against sensors. They were bright enough to cause sophisticated optical tracking devices to break lock long enough to prevent them from reacquiring a target before the rocket motor would burn out. Their obvious function would be to protect aircraft from incoming missiles. The round is basically a small air bag, again containing a small amount of high explosive surrounded with a noble gas. When detonated, it is tantamount to turning on the sun instantly.

One way to deploy these weapons is to fire them from air-defense missile launchers. In this case, the round would be propelled from the launcher and detonated in the vicinity of the intruding aircraft. Like the laser, the warning would be clear and unambiguous.

Electromagnetic Pulse Weapons

As the AGM-86 Air-Launched Cruise Missile winged across the Iraqi desert, it looked like any one of dozens of such missiles launched against key installations during Desert Storm. Targeted against the air-defense command center in Baghdad, it followed the computerized precision-guidance instructions, winding its way to within ten meters of the protruding antennae. There detonation occurred, releasing a huge electromagnetic pulse. With their sensitive electronics instantly fried, the Iraqi air-defense systems went dead.[11]

Though reported in *Defense News,* this attack never happened, but in the future it might. Electromagnetic pulse (EMP) weapons have been the Holy Grail of directed energy. The U.S. military has poured hundreds of millions of dollars into research and development of pulse-power weapons, and foreign countries have done likewise.[12,13] For decades, the developers of these systems have made promises about significant breakthroughs leading to the advent of a new dawn in warfare, but for the most part, the promises have fallen short.

The basic concept of EMP weapons is to generate one or more very intense pulses of electromagnetic power that penetrate equipment to degrade or destroy sensitive electronic circuitry. The concept is feasible and comparable to when a lightning bolt strikes in the immediate vicinity of computers, television, or other electronic equipment. This electromagnetic "attack" leaves equipment burned out or otherwise damaged as an electrical surge travels through the power cables and overloads computer terminals.

To damage or destroy electrical equipment, it is necessary to get the EMP into the device. This process is called coupling. There are two forms of coupling. They are known as front door and back door coupling. Front door coupling refers to energy that enters via an antenna or other path that is open to the outside and leads directly to the target device. If the operating frequency of a target system is known, the pulse can be tailored to do maximum damage. Back door coupling happens when a standing wave of energy is produced and travels to the target indirectly. The EMP may come through electrical cables, poorly shielded interfaces, or even holes in the walls of the system. Front doors are more desirable, as they allow the energy to flow unimpeded to the target. However, low-frequency EMP will couple well into wiring infrastructure, power-cables, and telephone lines. Once a transformer or shielding is breached, low voltages are sometimes sufficient to produce major damage to electrical equipment.

In warfare, EMP was first observed in high-altitude airburst nuclear weapons testing. Although it was a by-product, EMP damage to electronic equipment in a nuclear exchange was a major concern to the military, and extensive programs to harden weapons systems were started. However, shielding provides only limited protection and can increase the cost of the weapon considerably. Commercial computer equipment is very susceptible to damage from EMP. Therefore, hardening of the civilian infrastructure would be prohibitively expensive, and technologically advanced countries will remain vulnerable to EMP attacks.

Not all EMP is generated via nuclear explosions. Extensive work has been done on nonnuclear EMP. Approaches included explosively pumped flux-compression generators—circular explosives that fire inward, thus creating extremely high pressure and an EM pulse; explosive or propellant-driven magneto-hydrodynamic generators; and high-power microwave (HPM) devices, such as a virtual cathode oscillator, called a vircator. Other HPM devices under development include relativistic klystrons, magnetrons, and reflex triodes.[14]

In the United States, early explosive research and development was conducted by Max Fowler at Los Alamos National Laboratory. Later, Bob Reinovsky joined Fowler's efforts, and they were able to increase power output dramatically. Over time, flux-compression generators were capable of producing tens of megajoules, with peak power on the order of tens of terawatts. That's the equivalent of the amount of power put out by more than 1,000 nuclear reactors condensed into a single pulse. Precision-guided weapons have resolved one of the early conceptual problems. Because they permit pinpoint targeting, the area covered by the EMP, or footprint, which ensures that the target is in range, can be much smaller than originally thought. Therefore, the power requirements to damage a target are lower, reducing the amount of explosive and the size of the system. Once the technical challenges of producing these weapons are overcome, in the future an explosive-driven EMP will be packaged into weapons-delivery systems possibly as small as a 105-mm Howitzer round.

The Soviet Union also had major programs in development of EMP weapons. Although still debated, many in the U.S. intelligence and research communities believed the Soviet program was well ahead of U.S. developments at the end of the Cold War. Other NATO countries have also invested heavily in EMP weapons research. French, British, and German advances in the field have been substantial.

However, major problems of consistency of results have always arisen during the testing of developmental systems. Damage thresholds for similar components vary widely. Tests exposing identical parts to the same energy levels show that some are damaged, while others are not. In fact, some identical samples may not be damaged at energy levels more than an order of magnitude greater than pulses that destroy other samples. It has been a very frustrating experience for warfighters who want a dependable weapons system.

Another perennial issue is fratricide. Electromagnetic pulses do not discriminate between friend and foe. Therefore, delivery concepts have focused on mechanisms, such as missiles, that carry the warhead away from friendly areas and as close to the target as possible.

Advances will continue to be made in EMP weapons. Power output is going up steadily and refinements in the prototype systems constantly improve weapons capability. For law enforcement purposes, several laboratories are working on prototype devices that can stop vehicles. The Army Research Laboratory has demonstrated remote EMP weapons that have stopped many commercial vehicles, including both foreign and domestic models. Jaycor designed a weapon that could stop a vehicle by sending a charge from a pursuit vehicle directly into the one being chased. A variation of this technique was designed by David Patcholok of Non-Lethal Technologies, Inc. His system projects a tiny model car that speeds under the pursued vehicle. The antennae touch the metal frame of that car and allow direct discharge into the vehicle, killing the engine. Such systems will also have applicability for military peace support operations. While military targets are attractive to planners, the real threat is to civilian infrastructure: telecommunications, financial, transportation, and energy distribution systems. A massive EMP strike could cripple a nation temporarily. Recovery would be very expensive and probably done on the aggressor's terms.

Electronic Stun Guns

The Taser is fairly well known to the American public. This weapon was initially designed in the 1960s but was not accepted for use by the Los Angeles Police Department until 1980.[15] With it a police officer can quickly, and safely, incapacitate an aggressive individual. As seen in the Rodney King case, it can also be used to punish if not employed properly. However, contrary to the claims of King's lawyers, the Taser does not inflict burn marks or cause fatalities. In fact, LAPD reports that police officers use Taser over pepper spray at a rate of 5:1. While people doused with pepper spray are a problem for officers to transport to jail, Taser does not produce an odor or irritant.

One innovation was created by the manufacturers of Air Taser, in which they use compressed air to shoot the electrical darts at the aggressor. A laser sight allows accurate aiming at distances up to fifteen feet. This permits standoff distance for the user, but can be used on additional subjects in a handheld mode. This may be necessary in case of multiple assailants. In many states, this weapon has been declared a legal self-defense weapon.

Taser is a high-voltage, low-amperage weapon. Powered by a nine-volt battery, it delivers a 25,000-volt shock that causes loss of neuromuscular control. The affected person normally falls to the ground due to the inability to operate his or her legs. To ensure that the subject remains compliant, shocks may be readministered as necessary, but a single shock is usually enough!

Electronic stun guns offer an excellent alternative to lethal weapons, chemical spray, and clubs. They have applicability for police, military personnel on peace support operations, and for personal self-defense. Additional standoff distance would be desirable, and training is a key ingredient before anyone should be permitted to use them.

Light

A final system that deserves mention is that of simple bright lights. With improvement in batteries, handheld lights are now powerful enough to dazzle or temporarily blind a person. When the light hits the eye, reflexive closure occurs. Saturation of the vision cells may lead to the loss of ability to recognize contrasts. At night, when the pupil of the eye is more widely open, the effects are very strong. In face-to-face confrontations, these lights, rated at as much as 6 million candlepower, are sufficient to prevent an attacker from seeing the person holding the beam.[16] In fact, they are so bright that one can read a newspaper by their light a mile away. In confrontations, they can prevent an aimed shot from being fired toward the light. In addition to handheld lights, illuminating grenades are available for use in hostage-type situations. The grenades produce flameless chemical light that lasts for several minutes, temporarily blinding the criminal or terrorist while the situation is brought under control.

Strobe lights also have a part to play. These may be used to disorient people and inflict a temporary dazzling effect. Strobes may effectively be used as short-duration, point-defense mechanisms. Caution should be noted: some people are subject to epileptic seizures when exposed to strobe lights. There are a few people who have never displayed epileptic symptoms who may have a seizure triggered by these flickering lights. In fact, in December 1997, a Japanese animated television program, *Pocket Monsters,* accidentally triggered such a response, sending 700 people to the hospital. Reactions included vomiting blood, seizures, nausea, and, in a few cases, loss of consciousness.

Another innovative approach is to alternate the colors of the lights being used. Red light is at the low end of the visual spectrum, while blue is at a much higher frequency. Under normal circumstances, your eyes adjust to the light available. This concept, developed for law enforcement applications, is the simple irradiation with alternating red- and blue-colored lights. This alternation forces the eye to attempt to adapt rapidly from one end of the spectrum to the other. The result is conflicting messages to the brain, causing the individual to feel very confused and unsteady, and thus allowing the officer to take him or her into custody without physical injury. The officer can be protected from the severe changes in light by wearing sunglasses. Of course, countermeasures work both ways, but few criminals will be prepared for a light attack.

Novel Systems

Finally, it is possible that a unique EM weapon that can prevent guns from firing may already have been discovered. George Hathaway, an electrical engineer from Toronto, Canada, first told me about a garage inventor by the name of Sid Hurwich. Nearly fifteen years ago, George told me that Hurwich decided that something had to be done about the increasing number of robberies. Reportedly, he built a device that would not allow a pistol in the vicinity to fire. The operating principles were unclear, as demonstrations were conducted with the device fully covered from view.[17]

Were it not for a most remarkable and authenticated set of events, this story could be relegated to folklore. According to news sources, the Hurwich device was employed on the 3 July 1976 Israeli rescue operation at Entebbe, Uganda. The raid is legendary in the annals of special operations. During the raid, 103 hijack hostages were freed. The only Israeli casualty was the battalion commander, Yonatan Netanyahu, who was killed at the airport. What amazed all observers was that the Israeli airplanes had flown the entire distance without being detected, and very limited shooting occurred once they landed at the Entebbe airport. We cannot be certain that it was Hurwich's device that allowed the success of the raid. However, we do know that shortly following the incident, at Toronto's Beth Tzedec synagogue, Sid Hurwich was presented with the prestigious Protectors of the State of Israel award for his contributions to the raid.[18]

Hathaway talked with Hurwich about his invention several times. Hurwich indicated that while he had received the award, he had also been put on notice that he was not to discuss the operating mechanism with anyone ever again. To the best of our knowledge, he did not. Three years ago he took the secret to his grave. It was reported that his invention was not a highly sophisticated device, but merely a new application of old technology.

6

CHEMICAL OPTIONS

It's a non-lethal area neutralizer.
—*Michael Crichton,* The Lost World

SURRENDER OR WE'LL SLIME YOU read the headlines of more than one article about non-lethal weapons.[1] Due to these captions and the accompanying pictures, sticky foam and the concept of non-lethal have become almost synonymous. Sticky foam captures the imagination and seems to embody what the essence of non-lethal represents: an enemy or suspect attempting to flee but held fast by a gooey substance with the tenacity of Super Glue.

The reality is that sticky foam was developed as a potentially lethal weapon. Imagine this Hollywood script. Late at night a terrorist sneaks into a heavily guarded nuclear weapons storage area. In true James Bond fashion, the terrorist slips unnoticed by the dozing security officer (guards are usually sleeping, watching TV, or are otherwise inept in these scenarios), disables the multiple sensors (terrorists always have security systems designers on their team), bypasses countless locks (they have locksmiths too), and gains access to the warheads.

Even if this fictitious version of nuclear weapons proliferation were true, the terrorist would face the final challenge: sticky foam. Sensing an unauthorized intruder, the embedded sticky foam canisters would immerse the warhead and surrounding room with some of the most adhesive material in the world. Anyone caught by the foam would still be there in the morning. Worse, if they fell, they would not get up—*ever*! But then, lethal force is sanctioned to protect our strategic systems. That's right: Sticky foam was developed by Peter Rand, working at Sandia National Laboratory, when he was tasked to design a fail-safe, last-ditch system to protect our nuclear inventory.

Since recent interest in non-lethal weapons has increased, sticky foam has evolved into new uses and developed more methods of employment. Sandia technicians developed a backpack carrier and nozzles for a dispenser system so the foam can be deployed to the field. In fact, it was present, but not used, during United Shield. Sticky foam represents one example of existing technologies that have been adapted for new, non-lethal weapons roles. Thermoplastic resins used in chewing gum and combined with elastomers are the basis of these low-toxicity foams.[2,3] As the exceptionally adhesive version was quickly converted to a non-lethal purpose, there are two major complaints about the foam: its potential for accidental lethality, and its difficulty in removal. However, these objections to existing sticky foam can be engineered out.[4] While the initial version was so adhesive it would suffocate a person by getting into their respiratory system, future foams can be made less viscous. To remove the current sticky foam requires substantial time and effort using baby oil. Cleanup can be accelerated and made easier as part of an advanced development effort. The only remaining technical issue is for some agency to state the requirements to which the foam should be tailored. Of course the total system must be considered when deciding whether or not to adopt a new system. As seen in the photo section, the sticky foam dispenser is man-portable. However, the recharging unit, which operates with high-pressure nitrogen, weighs about 900 pounds. Each time the dispenser is fired, it must be refilled from this heavy unit that requires a vehicle to transport it. Preliminary field tests suggest that the best role for sticky foam may be in establishing barriers rather than restraining unruly individuals, as was first proposed.

A wide variety of non-lethal chemical options currently exists. Of course, even the word *chemical* evokes an emotional response, whether or not humans are affected. A number of antimatériel chemical weapons actually provide humanitarian options to the battlefield. Even some of the antipersonnel weapons available—ones that temporarily incapacitate—could reduce fatalities in future conflicts to both combatants and noncombatants alike. Political acceptance of new chemical options is the core issue. Opponents of *all* chemical weapons are faced with a conundrum. New materials are constantly being developed for a wide range of purposes. Some have potential military applications. It is unrealistic to believe they will only be employed for peaceful purposes. Possibly, they may save lives when used as weapons, so opponents of chemical weapons assume an illogical position and argue in favor of more killing and brutality. In this chapter, we will examine some of the technical options that are available, or could be in the near future. The technologies are hard to categorize, as there are exceptions, no matter how the field is divided. Two main categories are *antimatériel*, substances that affect the machines of war, and *antipersonnel*, substances that affect people.

Antimatériel Chemical Agents

There are so many antimatériel chemical agents that no single document could cover them all. To understand the breadth of these systems, I will address several in each of the following target categories: tires, engines, fuels and lubricants, tubing and seals, traction, optics, and other selected system components.

Tires are very attractive targets, and they are susceptible to chemical degradation. In many places around the world, tires constitute a critical, nonrenewable resource. That was true in Desert Storm, where many situations seemed ideal for an antimatériel attack. Largely due to the vast desert terrain, heavy vehicle travel was restricted to a few long roads. Supplies from outside Iraq could come only from Amman, Jordan, on a single, constricted road. Internally, the roads from Baghdad to Basra and beyond were equally limiting. The UN embargoes meant that tires could not be imported legally.

The operation, while feasible, never took place. While I held discussions about attacking these critical logistical targets with various Pentagon officials, it was clear that a new system could not be made ready, even at the prototype stage, before the ground offensive began. Frankly, I do not believe that developing any new system as you prepare for immediate attack is a good idea. Develop the weapons first, then test them and train with them before, not during, combat. The exercise did, at least, provide an understanding of both the problem and the possible solutions.

Superacids were the agent of choice for the degradation of tires. Based on combinations of fluorine compounds, these superacids are millions to billions of times more aggressive than hydrofluoric acid. Most will remember hydrofluoric acid from chemistry class—it was the one that was kept in wax containers because it could eat through glass. The counter-tire concept includes injecting a small amount of superacid into the tire. This may be accomplished with a small, four-pronged device known as a caltrop. First used by foot soldiers to equalize the awesome power of charging cavalry more than two millennia ago, caltrops always land with one point skyward and would pierce the hoofs of onrushing horses.

If a mere temporary flat tire is the objective, a simple caltrop consisting of hollow tubes that let the air escape, even from self-sealing tires, is sufficient. However, if permanent damage is necessary, then improved caltrops are required. The new, more sophisticated version acts like a syringe, inserting the disabling agent into the tire. After injection, a chemical reaction takes place in a matter of minutes, causing the tire to become brittle and to begin to delaminate. The intent of the attack is not to cause catastrophic disassembly, but, rather, rapid degradation leading to numerous flats and a breakdown of the land transportation system. It takes about a 20 percent degradation in strength before the tire begins to come apart. Instead of running 50,000 miles or more, on a set of tires, they self-delaminate in less than fifty miles.

The principal researcher in non-lethal application of superacids, Dr. Scott Kinkead, soon came upon a new, and even safer, approach to attacking tires. Catalytic depolymerization agents were just beginning to be explored by a group at the University of Florida. Not surprisingly, most of the work on tire chemistry had been focused on increasing the strength of polymer bonding, thus leading to longer tire life. Conversely, the new work was on agents that decreased the strength of those polymer bonds. A catalytic agent has the advantage that only a very small amount will create a large reaction. Therefore, a limited volume of these chemical agents would have a greater effect than the superacids we had previously proposed. Delivery could be in much smaller devices, but the effects would still be devastating. As supply-laden trucks rolled along, their tires would lose strength and come apart, leaving remnants strewn along the road.

There is another benefit to researching catalytic depolymerization agents. They solved a major peacetime problem: how to destroy old tires. The United States and many other nations are plagued with ever-growing mountains of used tires that are beyond retreading. Many other items of trash degrade naturally; tires do not. The method of dealing with this problem has been to assemble them in unsightly piles in out-of-the-way places. Unfortunately, spontaneous combustion sometimes ignites these repositories. Once tires start burning, they are nearly impossible to extinguish. Catalytic depolymerization agents could help control this monumental waste-management problem by reducing the tires to a powder that can be recycled. The carbon can be removed from the powder and used in another tire, while the remainder of the material can be placed in other products.

While attacking tires was initially considered, it was quickly determined that there are very few, if any, material substances that cannot be degraded by one form of superaggressive agent or another. Superacids, supercaustics, or aggressive solvating agents can be found that can degrade or destroy *any* target. These agents are so powerful that only a small amount is necessary to destroy the designated target. While no area coverage would likely be authorized, precision delivery by methods that ensure no contact with humans is feasible. These agents can be delivered as a liquid, spray, or vapor.

Tests have been conducted to demonstrate the effectiveness of some of these agents. To destroy computer board circuitry requires a milliliter or less. To eat through a quarter-inch of aluminum takes about twenty minutes with a device the size of a soft drink can. Glass optics can craze almost instantly when exposed to a vapor.

While almost every material can be attacked with an aggressive agent, there is no single universal solvent. In other words, no one chemical attacks everything. Also, chemical reactive laws apply, and the amount of chemical necessary to degrade a target may be too great to be practical. I have been asked about dissolving tanks with superacid. While it could be done, the amount of superacid required makes it totally impractical. However, by locating critical nodes, such as optics, computers, or seals, larger weapons systems become imminently vulnerable.

Of course, the mere mention of an agent as aggressive as superacids often produces loud, negative, emotional responses. With point delivery of such a munition, it could be a safe and reasonable option. Training, confidence, and logistics issues need to be resolved before superreactive agents are introduced into the field. They will be best suited for the unique requirements of highly skilled special operations forces assigned to destroy a critical subelement of a major system. With hand emplacement or advanced delivery systems such as intelligent minirobots, they will be able to strike designated targets without fear of collateral human casualties.

There is obviously concern about handling these very aggressive chemicals. One approach is to use a binary system. Less reactive chemical components are divided into separate containers and combined at the time of delivery. This adds a degree of safety, but does not eliminate concern. Later, we will discuss use of microrobots (millimeter size) in unique ways that would allow precision delivery of highly aggressive reagents.

Engines—automotive, aircraft, naval, or industrial—also make excellent targets. Attacks may be made that stop the engine at a critical moment in time, degrade functioning, or destroy the device altogether. Like humans, most engines breathe air. The air is taken in through a filter system, cleaned and forced into a combustion chamber where it mixes with fuel and ignites. Every step in this process is a point of vulnerability.

Filters are naturally vulnerable. It was failure of U.S. Navy Sea Stallion helicopter engine filters en route to the forward base called Desert One that brought an end to the attempted rescue of the American hostages at the embassy in Tehran in 1980. Flying across the Iranian desert, the helicopters encountered an unexpected sandstorm. The flying particles were so intense they completely filled the air-intake filters, leading to the loss of power that brought them down. Due to the absence of capability to fly everyone out of Iran, the entire mission had to be aborted. It was a significant defeat for both the American military and President Carter's administration.[5]

Like the desert sand, chemical filter-cloggers prevent air from passing through the air intake into the engine. Thin, long-chain polymer films that cover the external surface of the filter can be quickly deployed from a spraying device. Whereas most heavy-duty filters are designed to absorb large numbers of small particles, the long-chain polymers form a barrier across the outer surface and quickly choke off the air supply. Some tanks of the former Soviet Union had filter bypass mechanisms installed. This was to prevent incidents like the one that occurred at Desert One from happening at critical times in battle. However, the nature of the filter cloggers is such that ingestion directly into the engine causes them to destroy it entirely.

Another insidious approach is the introduction of extremely strong abrasive material into moving parts of an engine. Small, extremely strong ceramic or Carborundum particles can be distributed and ingested into an engine. The abrasives dramatically increase friction and grind away sensitive moving parts. They do not produce instantaneous results since they are designed to cause engine failure over a period of days. They would have military applicability in slowly developing situations and in situations where replacement parts are limited. The most effective way to introduce ceramics into an engine is by way of a metal fireball. Igniting certain metals, such as cerium oxide, produces extremely fine ceramic dust. The particles are so small they can penetrate filters and become interspersed throughout the moving parts.

Altering combustion is another method of stopping engines. Two approaches are viable: inhibiting combustion or overenhancing it. Through an accident in fighting a forest fire, it was learned that acetylene gas in the proximity of an engine will stop it from running—permanently. As the trees burn, they exude acetylene, which is highly combustible. When drawn into the engine, it causes the RPMs to rise very quickly. Hearing the change in engine noise, the operator must either climb on top of the engine or run like hell, as it will soon blow the pistons out the side of the block. Other gases that change the fuel-air ratio will have the opposite effect. These gases inhibit internal combustion and shut the engine down temporarily. However, once the vapors dissipate the engine can be restarted.

Research in this technology started with a lethal purpose in mind. There was always concern by NATO commanders about the number of tanks the Soviets could push forward simultaneously. Tanks rarely come alone; they usually invite many friends to the party. Even our technically superior tanks could only kill one at a time. The operational requirement was to have a method by which we could slow Soviet tanks down long enough for one U.S. tank to destroy multiple targets. The concept was to stall the tank in the open, then hit it with a HEAT round before the engine could be restarted. In military jargon, our tanks would be able *to service more targets.*

In addition to stoppage for short periods, long-term damage is desired in some cases. An unanticipated increase in power or heat can also stop an engine by causing it to burn up. One method is to have pyrophoric particles, such as cesium, drawn into a combustion chamber. There they would burn intensely and generate the heat necessary to destroy that engine permanently.[6]

Engine fuels are also vulnerable targets. Direct viscosification agents may be added to fuels, thus making it impossible for the fuel to be aerosolized in the engine. The gelled fuel would clog fuel pumps, pipes, or carburetors. The fuel does not change color, and contamination is very difficult to detect. Antidotes are possible; thus, the effects of an attack are reversible. Research into this technology has applicability in aircraft safety, as the fuel gels offer potential as flame retardants. The fuel in the tanks could be transported in a less combustible configuration, then have the retardant extracted prior to injection of the fuel into the engines. The process would diminish the probability of fire in the event of a crash.

One of the technology claims coming from the former Soviet Union includes a chemical that reacted so fast with petroleum that it could be injected into a flowing pipe and cause a blockage. Since the Soviets were plagued by frequent ruptures over very long and aging pipelines, this agent was reputedly developed so that repair crews could stop the flow of oil while pipes were fixed. Once the repair was complete, an antidote was injected upstream and the blockage dissolved. While it is known that polymers can cause fuels to gel, the claim seems a bit extreme. Nonetheless, work has been done to create effective agents that can make fuel temporarily unusable. The amount of the polymer agent required is still debated. To be militarily attractive, amounts will have to be far less than a 300:1 ratio of fuel to polymer by weight. Otherwise, the delivery requirements would be too great for the battlefield.

Obscuring vision is a viable method for inhibiting mobility. While crews of armored vehicles feel safe from small arms fire, their vision is greatly restricted. Small vision-blocks provide them with their only view of the outside world, allowing them to detect threats, and determine their ability to maneuver. It has been recommended for some time that obscuring vision would be desirable.[7] The chemical methods for blocking vision include simple techniques such as special bullets that splatter paint over the aperture, adhesive foams, and superacids that etch the optics. "Blinding" a tank may be one of the easiest ways to disable it temporarily.

Everyone has seen movies in which traction is affected by slippery substances used during chase scenes—oil on the road, marbles on a sidewalk, etc. The concept can be used in an antimatériel military role. Rather than degrading the machine, this approach alters the conditions on which it operates. During World War II, the French Resistance actually used slippery material on steep railroad tracks to prevent the Germans from operating their trains effectively. In recent years, materials scientists have developed new chemicals to reduce static friction. These are very important in making machines run more smoothly and with less internal wear. But there is also a military application. Superlubricants, such as Teflon spray or potassium soap, can be placed on the ground, making movement nearly impossible. Small capsules containing the superlubricants could be placed in the area of attack, then crushed by the target vehicles, greatly inhibiting mobility. Colloquially called *slick'ems,* they can be quite effective.

A new generation of antimatériel agents is currently under development and may resolve one of the major drawbacks to existing agents. Steve Scott at Sandia National Laboratory reports that they are working on *smart materials* that can alter chemical bonds on command. These chemicals activate under prescribed circumstances such as changes in temperature, electromagnetic signals, or change in pressure. One problem with super slick'ems and stick-'ems is that they are hard to clean up when it is time for friendly forces to enter the area. With smart materials there will be an ability to turn the effects on and off chemically when desired.[8]

Antipersonnel Chemical Agents

The military and law enforcement uses for chemical agents used against humans include crowd dispersal, subduing a criminal suspect, and area denial. In all cases, it is necessary that every possible measure be taken to ensure the recovery of the people attacked. Again, some of the technologies presented are controversial.

There is a long history associated with the use of chemical weapons. First known applications date over 2,000 years ago, when the Chinese used ground pepper to blind opposing troops temporarily. In 428 B.C., the Spartans used fumes from sulfur and pitch, and later "Greek Fire" was employed to suffocate enemy soldiers. The World War I applications are well known and have already been covered. The issues and many of the agents have been around for a long time and will not go away because of treaties.

Before addressing antipersonnel chemical agents, there is one topic that always arises and should be put to rest immediately. There is no *magic dust* or chemical dart that will instantly put people to sleep and then allow them to recover fully. Many movies have portrayed these mythical agents. In *Close Encounters of the Third Kind,* Steven Spielberg portrayed helicopters spraying an incapacitating gas on civilians climbing Devil's Tower. Then all who were exposed to this miraculous agent fell to the ground unconscious. Next they were picked up, moved, and recovered fully at another site. In other movies, darts are used to knock out guards, villains, and the like, usually in midsentence.

While there are fast-acting drugs, all pharmaceutical reactions with humans are based on many complex factors, including body weight and physical condition. The dose necessary for an average man may kill a child. A lighter dose would fail to incapacitate. Furthermore, terrorists long ago discovered the dead-man switch. Generally depicted as a person holding a grenade with the pin pulled, the dead-man switch is any device that activates when muscles are relaxed. No known drug produces an instantaneous catatonic state, i.e., keeping the muscles rigid while the body remains incapacitated.

The National Institute of Justice, in conjunction with several laboratories, has attempted to produce a safe and effective incapacitant as an alternative to the use of deadly force in stopping fleeing felons. This study was a direct response to the Supreme Court decision in the 1985 case *Tennessee v. Garner,* in which an unarmed adolescent was killed fleeing a nonviolent crime scene. It was undertaken with the best interests of both law enforcement officials and the public. Unfortunately, they were not successful in finding a suitable substance. While we don't have these magic weapons, and probably never will, there are many non-lethal alternatives that are viable.

Water is probably the most basic non-lethal antipersonnel weapon. High-pressure water cannons are effective and currently in the inventory of many police agencies around the world. By themselves, they are not controversial, but additives may change that. Dyes can be placed in the water so as to facilitate later identification of individuals who were present at an event. The dyes can either be visible or in the ultraviolet (UV) region. Using UV dyes could permit marking without the subject being aware that he or she may be easily spotted with a simple black-light device. The dyes are benign and are already used by clubs and dance halls to mark people who come and go at their events.

Riot control agents are found throughout the world. Though civilians tend to think of them collectively as tear gas, these agents can affect many human functions, causing tearing, nausea, vomiting, and sometimes pain in the form of a burning sensation. Due to their long chemical names, riot control agents are simply called by two letters, such as CN, CS, CM, or OC.[9] These agents may be sprayed as a vapor or mixed with water, and while they are toxic, they are not believed to cause a risk to human health under these limited applications of short endurance. Symptoms usually disappear within twenty-four hours of exposure. Irritants may be released through smoke grenades or in aerosol form and affect the eyes and mucous membranes of the respiratory tract. Increased concentrations may produce temporary blindness and conjunctivitis.

Use of these agents should be in the open, or in an area that can be ventilated. Concentrations at very high levels may cause toxic pulmonary edema, which can be life-threatening. Those most at risk are young children, the elderly, and people with respiratory illnesses.[10] While risk is involved, these agents offer a viable alternative to lethal weapons. It will be left to on-scene decision makers, who should have received extensive training, to determine when it is appropriate to employ riot control agents.

Foam technology offers many new options and advantages. One advantage is the expansion ratio between the compressed material and the foam generated. The sticky foams mentioned earlier expand about 50:1 by volume. This means that for each liter of ingredients in the backpack, about fifty liters are dispersed. The foam has an adhesive tensile strength about an order of magnitude greater than common sticky materials such as molasses.[11] A little goes a very long way toward inhibiting mobility.

Besides sticky foam, there are several other varieties. Aqueous, or water-based, foams expand at a ratio of about 1,000:1, thus offering attractive advantages of safety and volume. They, too, have been around for a long time. First developed in the 1920s to fight coal-mine fires in England, they also have been used for dust suppression.[12] The foam is made up of a 2 to 5 percent solution of Steol CA-330 surfactant, a substance used extensively in the cosmetics and hair-care industries. Their great expansion means that most of the foam is actually filled with air and has a consistency much like soap suds. Therefore, although persons engulfed in the foam may feel uncomfortable, they are in no danger of suffocation, but they are likely to become disoriented very quickly, however. When water-based foams are used in prison riots, cells can be quickly inundated, preventing coordinated attacks. The aqueous foams are also good carriers for irritants such as OC, thus enhancing the breakup of any disturbance.

Large foam generators exist that can spew forth hundreds of gallons per minute. In an external test, a large amount of aqueous foam was sprayed over a road. Although the initial depth was only about eighteen inches, when a vehicle was driven into it, the foam rolled up, covering the windshield in seconds. The vehicle was forced to stop almost immediately. Aqueous foams can also be used to cover the use of other weapons. For instance, in addition to making a road slick, they can hide caltrops that puncture the tires of vehicles that attempt to penetrate the foam barrier.

In other tests, people have been placed in enclosed areas and submerged in foam. They quickly became disoriented and could not find their way to a door only a few feet away. These tests demonstrated that aqueous foams definitely have a place in our non-lethal arsenal for both breaking up internal disturbances and external crowd control.

Malodorous chemicals, colloquially known as *stink bombs,* are useful for area denial or for rooting out suspects. Some agents are powerful enough to induce gagging or vomiting. Others, such as putrecine or cadavercine, issue odors to which humans are naturally averse. Not forming an impenetrable barrier, these obnoxious smells can prevent an area from being occupied or used for any length of time. Of course, when sprayed on people, these agents will usually cause them to leave and get cleaned up. A practical application may be in circumstances when it is necessary to extricate our forces from otherwise friendly crowds.

There is even a role for sex in non-lethal warfare. This entails the use of pheromones, a chemical substance that evokes sexual behavior in others of the same species. In the past year, advertisements for pheromones for sexual attraction between humans have started appearing in magazines. The non-lethal role is a bit different, but still based on sex—insect sex, that is. The concept is to spray pheromones that attract a variety of creepy crawlers into areas that we do not want people to inhabit. Imagine trying to sleep or work in an area that is attracting every ant, cockroach, or spider for miles around. Infestation will probably exceed an adversary's capability to fumigate. With both odors and pheromones, decontamination is a problem. They should only be used when long-term dehabitation is intended.

Pyrophoric agents were previously mentioned as a method to produce damage to engines. They may play another role. In Afghanistan, the Soviet forces used pyrophoric tars as a method of area denial. These tars are placed on the ground and remain dormant until they are disturbed. Once the sealant breaks from pressure, air combines with the pyrophoric agent and flames erupt. Unfortunately, if a person touches the tar, given the adhesive qualities of that substance, severe injury could occur. However, with warning, pyrophoric agents can replace explosive mines as an effective barrier.

Several types of drugs may be used as non-lethal weapons. While extremely controversial, they are included here by way of acknowledgment, not endorsement. Soporifics, or sleep-inducing agents, are one class of drugs that is frequently mentioned. Barbiturates, benzodiazepine derivatives, diphenhydramine, and the infamous "knockout drop," chloral hydrate, are examples of soporific agents. The basic concept is far from new and can even be found in children's literature. Remember the mysterious poppy fields in *The Wizard of Oz?* Of course, in real life, administering the drugs and controlling dosage is very difficult—a point not addressed in the fictional accounts. Analgesics, such as ether or nitrous oxide, are candidates. It is claimed that the Soviet military actually developed an ozone cannon to induce drowsiness in opposing forces.[13]

Probably the most controversial of all antipersonnel measures are psychoactive drugs. Reportedly, more than 130 pharmaceuticals, such as sodium pentothal, scopolomine, and the famous hallucinogenic drug lysergic acid diethylamide, better known as LSD, were tested extensively by the Army and the CIA in the 1950s and 1960s in programs with code names like MKULTRA, MKDELTA, Artichoke, and Bluebird. While they proved effective in disorienting people, there were disastrous side effects, including flashbacks, psychotic breakdowns, and, in one infamous case, suicide.[14] In discussions with operatives who did not experience the above consequences of those programs, the consideration of using psychopharmaceuticals has arisen. For those who remember those consequences, renewing such programs is unthinkable.

Finally, should you need to disable a ravenous and cunning velociraptor, such as those made infamous in Michael Crichton's *Jurassic Park,* a miraculous chemical, metacholine, is reported to paralyze "all life forms for up to three minutes." Unfortunately, this universal, fast-acting, paralytic agent has yet to be invented. Based on physiological constraints, it probably never will be. However, we will undoubtedly keep looking for it.[15]

7

PHYSICAL RESTRAINTS

Those who are ready to die can be killed;
those intent on living can be captured.
—*Sun Tzu*

The infantry squad crept silently forward to the edge of the clearing. Their mission was to secure a small promontory and prepare to defend it against a landing party. Before them lay about 100 meters of open terrain. Spotting movement from the position they were to occupy, their only choice was to assault across this space as quickly as possible. The trained and experienced squad obeyed silent commands and prepared to attack. On command, a base of fire raked the tree line, and the attacking squad rose as one and rushed forward.

The first man fell twenty meters into the charge. Others followed, stumbling and falling down as if struck by an invisible force. This was a training exercise, and no withering live fire produced the casualties. Instead, it was a simple net concealed in the knee-high grass covering the field that befuddled the infantrymen. Although each man regained his footing, he would take no more that a few steps before tripping again.

When the films were reviewed, the clumsy antics of this skilled infantry squad were more reminiscent of the Keystone Cops of yesteryear than a professional military unit.[1] It again demonstrated an ambiguous element of weapons that cause temporary delay. Covered by machine guns and rifles, the nets would serve to increase the lethality in the kill zone. The troops would be exposed to enemy fire for a prolonged period of time—even if they attempted to retreat. However, the same nets could provide a non-lethal barrier, preventing casualties on both sides.

Dependable, simple, and effective, the technology that felled this squad is as old as warfare itself. One of the favorite options for arming Roman gladiators was the net and trident. In Stanley Kubrick's film classic *Spartacus,* we saw how effective the net could be in trapping opponents armed with swords and shields. The range was limited to the throwing distance of the gladiator, but normally the net was reserved for snaring the opponent at arm's length. Two basic changes have taken place over the centuries. One is the improvement of the materials used in the netting. High-strength, lightweight, synthetic fibers have replaced older hemp or manila ropes. Additionally, in recent years, advances have been made in projection techniques for both distance and accuracy.

In the 1996 film *Escape from L.A.,* Snake Plisskin (Kurt Russell), an ex–Special Forces lieutenant turned talented and despised criminal, is sent to the postcataclysmic island of Los Angeles to rescue the President's daughter, who has reportedly been kidnapped and held hostage.[2] Spotted during his arrival in a one-man minisubmarine, Snake is relentlessly pursued by many villains and eventually forced into a small, open area. Though an experienced street fighter, he is quickly captured; snared and disabled by a device known as a Netgun.

A riflelike shoulder weapon, the Netgun fires a blank cartridge that simultaneously projects four weighted, padded balls deploying an eight-foot-square net. Accurate for distances up to forty-five feet, it is ideal for capturing a single individual in an open environment. It provides officers adequate standoff distance for those armed with clubs or knives. The tensile strength of the four-inch nylon mesh is 200 pounds, making it too strong to be cut quickly as the suspect attempts to free himself or herself.[3]

The projection concept is not new nor restricted to humans. Nets and snares have been employed from antiquity on animals. Initially, they were used to capture them for food. More recently, they have been employed in preservation activities when animals have become injured or need to be relocated. Wounded animals can be very dangerous; thus, nets allow them to be brought under control without injury to the rescuers. However, liability issues dictate increased safety requirements when applying nets in a law enforcement situation. For instance, even if one of the projectiles hits a suspect in the head, the padding and low launch speed ensure that no serious injury will occur.

Larger nets have been designed for use at longer ranges and for targets that include multiple numbers of people. Under the direction of Arnis Mangolds, Foster-Miller, Inc., has developed a number of net alternatives for DARPA and the U.S. Army Armaments Research and Engineering Development Command (ARDEC). They have both antipersonnel and antimatériel weapons applications. In addition to nets designed only to ensnare people or things, other military activities may be conducted with advanced technical adaptations of the basic system.[4]

One such operation is clearing minefields. By lacing the nets with explosive cord—known to demolition experts as "det-cord"—they can be used effectively for this mission. The net is mechanically dispersed over the suspected minefield and detonated. Pressure from the explosion will set off most of the mines.

Another advanced application is as a countermeasure for incoming missiles. Foster-Miller's concept, called *Birdcatcher,* would sense the incoming missile, then project the det-cord constructed net in front of it. The net deploys, makes contact, and collapses on the inbound munition, then detonates and destroys it before impact on the intended target.

Dispensers for the nets come in a variety of munitions, including 2.75-inch rockets, MK65 bombs, and a small version, a 40-mm munition for ballistic deployment, compatible with standard M203 grenade launchers. It is this 40-mm munition that provides new capabilities without requiring another weapon to be carried by infantry units. The net size in this small round is limited to about an eight-foot diameter and is effective for ranges up to 90 feet.

Foster-Miller has also designed larger antipersonnel nets that can be deployed up to 100 feet and wrap three or four people inside. Applying innovation, they also developed a number of additives that give these simple systems new capabilities. To increase the time for people to free themselves, a sticky substance, based on the chemicals discussed in the previous chapter, may be placed on the strands. However, improvements were made that will increase acceptability, especially for law enforcement applications. These include water-based cleanup measures that allow removal in a matter of a few minutes, as opposed to hours for the sticky foam. They also developed a unique coating for the sticky substance that permits packaging without self-adherence. Net size can vary and may be designed to snare one person, or several at a time, if necessary, when confronted with a crowd.

Combining nets with electromagnetic technology provides another antipersonnel option. Thin wires can be carefully woven along the nylon fibers. Connected to a battery, a substantial shock can be given to anyone caught in the net or touching it. The shock can be delivered in a preset mode at a designated voltage and time duration. More advanced is a "smart" mode that senses struggle and administers a shock only when the suspect attempts to escape.

One of the funded non-lethal weapons projects is a vinyl net capable of stopping cars. The system must be preemplaced but is useful at checkpoint or border crossings. Called *Speedbump,* the system is designed to stop a 5,100-pound vehicle traveling at up to sixty miles per hour. The vehicle will be stopped within 200 feet without serious injury to passengers.

Nylon entanglements may also be employed during waterborne operations. Drug dealers frequently use very fast "cigar boats" to transport their illegal merchandise from container ships or after recovering an open-water air drop. If necessary, they have the horsepower to avoid capture. Reaching speeds up to 160 miles per hour across the open seas, many have the ability to outrun police or Coast Guard patrols. Operating under U.S. laws, the Coast Guard has zero tolerance for collateral casualties; thus, their lethal firepower cannot be brought to bear on fleeing boats.

A similar situation existed in Hong Kong before it reconverted to control of China, with drug smugglers evading British patrol boats and other local authorities. One method of stopping them was to dispense long rolls of plastic fibers on the surface ahead of the boats. Once the boats hit the fibers, they quickly became so entangled in their propellers or engine intakes that they would stall and become dead in the water. Simple countermeasures, such as cutting the netting to free the engine, did not prove effective. Placement of the netting is an important issue. In order to get ahead of the boats, an airborne platform, a light plane or helicopter, would usually work in conjunction with the high-speed police patrol boats. By dropping the lines a short distance in front of the fleeing boat, the smugglers were not afforded time to maneuver out of the path.[5]

While this method is effective, there is a downside. The light nylon fibers will continue to float on the surface. Unless they are retrieved, they will be a hazard to legitimate shipping and wildlife. The most vocal complaints will undoubtedly come from environmentalists concerned about dolphins, fish, or birds that might become tangled as they do in fishing nets. However, a simple resolution to the problem is to attach brightly colored markers, thus facilitating the recovery effort.

Barrier technology may also be used on land. Instead of nylon filament, lightweight metal wires are employed. Developed by ARDEC during the Vietnam War, handheld barbed-wire dispensers were designed to allow the military to establish an antipersonnel barrier quickly. The dispenser relied on energy from the coiled wire to project the entire length of wire at a range of fifty to eighty feet. In about eight seconds, a 450-foot fence could be emplaced from a canister only 3.6 inches in diameter and 10.25 inches long.[6] A larger system was developed for antimatériel missions. Each submunition contains multiple spools of concertina wire. The razor-sharp barbs are designed to shred tires, entangle drive shafts, and also to keep people at bay.[7]

Several studies have explored the use of entanglements for aircraft. In the NATO AGARD study on deterring noncooperative aircraft, we concluded that interfering with the pilot' s ability to control the aircraft during flight is likely to be lethal.[8] As a non-lethal alternative, it was discounted. However, viable concepts for temporarily controlling airspace with entanglements do exist. Clearly marked, lightweight nets or strands may be suspended over the area in which low-level flight is to be precluded. The barrier material may be dropped from above, released on lighter-than-air balloons, or even tethered to prevent drifting. The hang time of the material precludes sensible pilots from flying through that airspace until the barrier has been cleared out.

In addition to airborne entanglements, other drag-induction measures for controlling noncooperative aircraft were discussed by the AGARD study group. One technically feasible concept was an air-to-air harpoon with a small drag parachute attached. The drag device would cause the pilot to attempt to land the aircraft as soon as possible. Advanced versions recommended remote control of the drag device so that once compliance was indicated, the device could be deflated or disconnected from the harpoon. Again, the pilots in the study group indicated that changing the aerodynamics of the aircraft while in flight was too dangerous to attempt.

One of the most advanced antivehicle technologies was developed for DARPA by Foster-Miller in a classified program code-named *Cover.* Now declassified and called *Silver Shroud,* it is a ballistically deployed polymer film that literally wraps up a targeted vehicle. The aluminum P_4 coating is amazingly thin—0.0005 inch, to be exact. It can deploy over an area of 1,960 square feet in less than a second. As shown in the photograph section, the vehicle is completely enveloped. With vision totally obscured, an operator would have no choice but to stop. The polymer film adheres to the vehicle and becomes binding. When the coating has been applied to tanks, the turrets will become stuck if they are turned, thus further disabling the vehicle. Given the extremely thin films involved, a coverage for a large area can be packed into a relatively small munition or submunition.[9]

In recent years there has been a great deal of controversy concerning land mines. Millions of antipersonnel explosive mines litter old battlefields around the world, and every year, thousands of people, often children, detonate them. Some are killed in the explosions. Many more lose limbs or are otherwise maimed when they accidentally encounter the mines, which have been indiscriminately strewn about the countryside. When properly employed, mines are a useful tool of warfare. They are used to protect exposed areas that cannot be defended by troops. They are also used to slow down the enemy's movement and to canalize them into zones that are covered by heavy fire—all legitimate tactics. However, these minefields are to be visibly marked, and the pattern of the mines recorded so that the fields may be disarmed at a later date.

Unfortunately, in most of the conflicts around the world, mines have been placed in farmers' fields, in wooded areas, and in cities without regard for identification or registration The locations of the mines are then forgotten. That is why unsuspecting people have become the predominant casualties long after the war has passed by. The resultant number of injuries to innocent civilians has been so great as to cause an international public outcry. The United States, along with many other countries, is now engaged in serious efforts to ban the manufacture and use of such mines.

Net technology may be the next step in providing a safe alternative to mines. At the current cutting edge of non-lethal weapons are mines that deploy nets rather than shrapnel. Foster-Miller is now testing a claymore version as well as a model of the "Bouncing Betty" of World War II fame. Claymore mines are well known to Vietnam-era soldiers. They were thin explosive devices that dispersed hundreds of small BBs, like a large shotgun shell. Claymores were used to initiate ambushes, and as a defense against swarms of enemy troops. The new version deploys a sticky net against the intruders. The Bouncing Betty would pop up into the air and explode when it was about waist high. The non-lethal mine jumps up and shoots out a net.[10] Both styles will assist in accomplishing the usual mission of minefields—to slow down or contain an enemy—without the deadly consequences of high explosives. They will not solve the entire problem but are a step in the right direction.

In addition to previously constructed entanglements, it is possible to create high-strength fibers as they are needed. The concept is similar to the comic strip web-crawler *Spiderman,* famous for creating high-strength strands from liquid containers in his suit, from which he freely swings to, or from, danger. The larger weapons system has a simple binary assemblage containing hexamethylenediamine dissolved in liquid CO_2, and adipic acid in pressurized H_2O. When employed, the two chemicals are united under pressure, spewing forth a thin, long-chain polymeric fiber that is projected above the target. By altering nozzle design, multiple filaments can be generated. Range is dependent on pressure, but distances greater than 100 feet are attainable. By slight movement or rotation of the nozzle, a netlike entanglement is instantly formed. This chemical system has an advantage over ready-formed nets as it can be adjusted to the specific area most in need of targeting. It would make an excellent, quickly deployed antipersonnel barrier.[11]

While pursuit of fleeing vehicles is one of the most difficult problems in modern law enforcement, a simple barrier technology, spikes, has been developed to assist in stopping automobiles. The basic problem with spikes that puncture tires is that they puncture *any* tire. The innovation that made this approach viable was to put the spikes on a rotating strip. As the car being chased approaches, the police officer twists a knob and the spikes point up. As the car passes over, the spikes detach and remain embedded in the tire. As soon as that car clears the strip, the remaining spikes are retracted, allowing pursuing police cars to pass safely. Simple and effective, these systems are available in many police agencies today. A major drawback is that the police car with the spiked strips must get in front of the fleeing car. On an open highway, this is relatively easy to do, but zigzagging through city streets is another matter.

Some improvements have been made to spikes. One developed by PMG Manufacturing allows police to activate the spikes remotely, via a radio signal. This permits the officer to be farther away from the strip when the spikes are moved into the up position. They can also retract them before the pursuing police car becomes a victim.

Another innovation permits spikes to do more damage than just release air from the tire. During numerous high-speed chases, criminals have continued to drive on the wheel rims after the tires have been deflated. While this is extremely dangerous, the fear of being caught too frequently overrides common sense. Therefore, in a more aggressive form, explosives may be added to the spikes. Once detonated, they will prevent the vehicle from driving on the rims.

For military operations, caltrops can be employed. These are four-pointed scatterable spikes designed so that when dispersed, one point is always straight up. They can be quickly deployed but are not retractable. Lieutenant Sid Heal of the Los Angeles Sheriff' s Office added an innovation that prevents the caltrops from being easily spotted: Instead of the normal metal configuration, he designed a spike made predominantly from clear plastic.

Control of suspects while being transported to jail is a serious concern for law enforcement officers. Irate or emotionally disturbed prisoners, or those still high on drugs, are particularly difficult to handle. Even when handcuffed, they have been known to do extensive damage to police cars. In a few instances, they even have been able to kick out rear windows.

One simple solution was designed by Donna Marts at Idaho National Engineering Laboratory: air bags.[12] These are not the air bags you see displayed in automobile ads on television. Car air bags are driven out with tremendous force—up to 250 miles per hour—to counter the sensed impact and inflate in about twenty milliseconds after impact. They then quickly deflate so the driver can regain control of the vehicle. The air bag restraints for law enforcement are controlled by the officer in the front and are filled much more slowly, through a small fan. Within a few seconds, they gradually but effectively push the suspect against the seat, thus preventing the suspect from kicking against the front seat, doors, or rear windows. The material is a heavy canvas, not the polyurethane film found in auto air bags. It is strong enough to resist biting or scratching with sharp fingernails as the suspect attempts to free him- or herself. Sections of the air bag are air-permeable, minimizing the risk of suffocation. If the subject behaves, the officer can deflate the air bag. Larger versions have been made for use in small jail cells. Deployed from the ceiling, the air bag forces the prisoner against the wall and facilitates his or her extraction from the cell.

8

LOW KINETIC IMPACT

They thank you when they wake up and they're not dead.
—*Officer David McArthur, Las Vegas Police SWAT Team*

At 8:53 P.M., the dispatcher sent the 38 signal over the police radio.[1] *Domestic disturbance—it is one of the most common calls to which police respond. It is also one of the most problematic. As Police Officer David Jenkins arrived at the scene, a tearful woman came running toward the car. Doris Williams reported that her live-in boyfriend, Tom Powell, had come home drunk about an hour earlier and started a fight. As screaming escalated to hitting and throwing things, Doris had managed to barricade herself in the bedroom and dial 911. Minutes later, when an operator finally answered, Doris pleaded for the police to come quickly.*[2] *Against instructions from the operator, she hung up the phone and taunted Tom by telling him police were on the way. That really set him off! In a rage, he broke down the bedroom door, grabbed her by the hair, and threw her with sufficient force that she exited through the still-locked screen door.*

Emboldened by the presence of Officer Jenkins, Doris, herself half snockered, began to scream obscenities at Tom, who returned the pleasantries, upping the ante. As more police officers arrived, they pulled Doris to a police car, put her in the back seat, and locked the door. Under the recent domestic abuse laws, an arrest was imminent. The officers asked Tom to come out of the house so they could talk. Tom yelled back and told them to get off his property. It was his house and Doris was just a temporary visitor, he claimed. In his mind, the police had no business being involved in the altercation. Finally, Tom, obviously still drunk, came roaring out swinging an aluminum baseball bat. At six four and 230 pounds, he was a formidable foe. As he rushed at Jenkins, the nearest of the officers, two 12-gauge shotguns blazed. The impact of the rounds on his chest lifted Tom off his feet and literally knocked him on his butt. Certain that he was dying, Tom dropped the bat, clutched his chest, and began crying for help. It would be a few minutes before he realized that he had been shot with bean-bag projectiles, not buckshot. He would be bruised and sore for several days. But now he was alive and under arrest.

Not long ago, it would have been .38-caliber or 9-mm rounds that would have fatally pierced Tom's chest. Now there are families of non-lethal rounds that provide an alternative when an officer's life is not in jeopardy. These rounds have also reduced another unfortunate police incident, so common that it has assumed the name *suicide by cop*. Such circumstances arise when an individual decides to die, but either does not want to use his or her own hand or is afraid he might lose courage or botch the job. An increasing number of people, even children, are precipitating standoffs, threatening to shoot police or bystanders, then exposing themselves to fire. A tragic example is the 911 call placed by sixteen-year-old Julie Marie Meade on 22 November 1996. During that call, Julie outlined her bloody demise to the Prince Georges County, Maryland, police dispatcher. When the police arrived she confronted them brandishing a pistol. When she refused to obey orders to drop the weapon, the police did exactly what she wanted and shot her to death. On closer examination the police found that Julie Marie had been waving only an air pistol, but the damage had been done.[3]

There are many varieties of non-lethal rounds now available. The first to be used were baton rounds made of teak wood. In the 1960s they were employed against striking workers and anti-British rioters in the then Crown Colony of Hong Kong.[4] Rather than being fired directly at the rioters, the rounds were skipped off the ground so they would strike the legs and avoid more serious injury. Because of this technique of firing, the wooden rounds acquired the nickname "knee-knockers." The multiple-baton round commonly contained five blocks and was first used in the United States against student rioters at the University of California–Berkeley in 1971.[5]

By 1970, the British developed rubber and plastic bullets for use on the streets of west Belfast in Northern Ireland. Fired from a gas-grenade launcher, a long baton was thrown into a crowd quite effectively. It proved to be safe when striking the lower body area or limbs. Packing quite a wallop, the rounds leave the barrel at about 200 miles per hour. Rules of engagement stated that they were not intended to be fired into a crowd at a range of less than twenty-five yards.

These rounds became very popular with the British Army and Royal Ulster Constabulary, but they were hated by the Irish—so much so that they made their way into contemporary folk songs. The words to one song affectionately go, "Take your rubber bullets and shove them up your ass." Between 1970 and 1974, more than 55,000 rubber bullets had been fired in Northern Ireland. However, as with later rounds, if the subject was struck in a vulnerable spot, serious injury or death could occur.

The Royal Small Arms Factory of Enfield Lock went on to develop the Anti-Riot Weapon Enfield, or ARWEN. A single-shot weapon was made first, followed by one that fired five shots. The ARWEN 37, a five-shot, 37-mm carbine with a rotating cylinder, fired 2.7-ounce, 4-inch plastic rounds up to 100 meters. It was adopted by the United Kingdom Army in 1979 and the Los Angeles Sheriff s Department in 1985.

In 1986, the Sheriff's Office used it to put down a major uprising in the Men's Central Jail. Limited use was made in the 1992 L.A. riots in an attempt to stop looters. Since then, many police departments have adopted and used baton rounds against disturbed individuals or for crowd control. The relatively high cost (about $1,500 per weapon) has limited the number purchased by small departments.

Responding to the Intifada, and the years of unrest and violence on the West Bank and Gaza Strip, the Israelis have also developed a number of low kinetic-energy (KE) weapons and munitions. For more than a decade, these systems have been available and used by both the Israeli Defense Forces and the Israeli Police.[6] Some of their systems, such as the MA/RA 83 and MA/RA 88, are designed to attach to a standard M16 rifle or any other rifle with a NATO 22-mm flash suppressor. The containers can be fitted on the rifle in under six seconds and are fired using a 5.56-mm ballistic cartridge. This is a major advantage, as the shooter always has lethal firepower immediately available. The containers weigh about one-half pound and hold fifteen rubber cylinders. Each rubber ball weighs fifteen to seventeen grams and impacts safely with between thirty and thirty-six joules of energy.[7] The manufacturer, TAAS-Israel Industries, puts the operational range at from zero to eighty meters depending on the system used.

While the fifteen-ball munitions are good for crowds, TAAS-Israel has also designed rounds with fewer but heavier submunitions, designed to incapacitate a targeted individual. Wrapped in plastic, the rubber cylinders separate at three-to-five meters' range. The rubber matrix is filled with steel particles and is likely to cause bruising but not break bones.

For larger crowds, they have developed a cartridge containing seventy small balls, nineteen millimeters in diameter. The cartridge weighs about seven pounds and requires an independent weapon. A vehicle-mounted, multiple-cartridge version is available that fires up to 1,400 balls in a sequence determined by the operator.

Weapons developers in the United States have been busily making a variety of non-lethal low-KE munitions. By one recent count, fifty-nine different rounds and grenades are currently available, with new versions quickly following them to market.[8] Many of these rounds are very similar, but the number now available provides a feel for the market, as perceived by manufacturers. There are handheld projectors and rounds for conventional weapons—most frequently a twelve-gauge shotgun, gas-launch guns, and grenades. While some of these munitions are purely low KE, others have incorporated other technologies, such as light, gas, and acoustic effects.

Jamie Cuadros and his firm Arts and Engineering have been one of the leaders in American development of low-KE rounds.[9] He has conducted extensive testing of various types of munitions, including several of his own bean-bag rounds. Bean-bag rounds contain a number of small canvas bags, each filled with a small amount of lead shot. These bags spread the impact over a broader area, thus limiting the physical damage.

Another American firm, Accuracy Systems Ordinance Corp., of Phoenix, has developed a family of munitions designed to limit collateral casualties in terrorist or hostage situations. The owner, Charles Byers, has appropriately named them SPLLAT, for Special Purpose Low Lethality Anti-Terrorist shells. Among the unique designs is a round that can blast a lock from a door without danger of penetrating into the room beyond. This is accomplished with a frangible metal/ceramic slug that disintegrates in the process of destroying the lock. Even safer is a round without the ceramic binder. This round is employed at contact with the muzzle. Once fired, the lock is punctured, while the round produces only a cloud of fine metal dust. Also available are diversionary rounds that produce a tremendous blast while dispensing brilliant sparklets.[10] A photograph of these munitions being fired is included.

An example of a handheld device is a Baton Ball.[11] Developed in the United Kingdom, the 40-mm ball weighs about seven ounces, is projected by a blank cartridge, and may incapacitate for more than an hour. Several munitions mix irritants, such as OC or CS, with rubber and steel balls. STINGBALL is one of those, with 100 soft rubber marble-sized balls that both strike assailants and release CS.[12] The bean-bag rounds, the small canvas packets with lead shot inside, can also induce stinging and may be combined with other materials. Dye can be added to these rounds and used later to identify individuals who were hit, clearly indicating they were involved in the disturbance. To increase velocity and accuracy, 12-gauge rubber sabot projectiles have been developed.

STINGBALL grenades have been proven effective in breaking up prison riots. Once the prisoners learned about rounds being fired from shotguns, they employed mattresses as shields, but the grenades are fired over the crowd and disperse in all directions. Therefore it is impossible for rioters to use simple materials in self-protection. Once the STINGBALLs are fired, the disturbance is quickly quelled.

Another innovation is the ring airfoil grenade (RAG) round. The RAG requires a special launcher that fits onto the standard M16 rifle. Uniquely, the ring projectiles are launched and spin at a rate of 5,000 revolutions per minute for gyroscopic stability.[13] As it spins, the ring expands. Thus, when the ring hits the target, the impact is spread over a larger area, minimizing the potential for undesired serious injury.

All low-KE rounds come with a warning about proper use and acknowledge the possibility of serious injury or death. As we have mentioned, nowhere has that issue been more emotional than in Northern Ireland, where seventeen people have been killed accidentally by plastic or rubber bullets.

The Irish Catholics have borne the brunt of these injuries and claim the soldiers have intentionally abused these munitions by firing at ranges closer than authorized. If you examine the statistics, however, the actual fatality rate is extremely low. The seventeen fatalities occurred over a period of twenty years, during which more than 110,000 low-KE rounds were fired. That is less than one per year and about 1 per 6,500 rounds fired. Therefore, if the low-KE rounds are only used during periods of violence, the fatality rate would be much higher if live rounds were used. Unfortunately, of those killed, a number were teenaged children, making the issue more volatile.[14]

As a primary rule, force should not be used unless some level of casualties is acceptable when weighed against alternatives. In Northern Ireland, those alternatives included not responding until violence got out of hand, or using other weapons, including traditional light infantry systems. In either case, it is highly probable that the fatality rates would be much higher.

Low-KE munitions injuries have been compared to baseball impact. It is known that a baseball hitting a child in the chest at eighty miles per hour is the threshold for death.[15] If struck in the unprotected sternal area of the chest, the heart, liver, and spleen are at risk. Damage may range from contusion of the heart to rupture of the heart muscle and separation of the arteries, leading to fatal bleeding. The same impact on large muscle group areas, while painful, will do little physical damage. Obviously, the head is also vulnerable. Several deaths in Northern Ireland were attributed to plastic rounds fired at close range and striking the head. While accidents may always occur, properly acquiring targets is a training and discipline issue.

Additional testing is being done to determine the extent of physical injury from various types of non-lethal munitions. Much of the data are extrapolated from studies done with compression tests involving baseballs delivered at precise speeds. The results indicate that if rounds are delivered in a proper manner, the probability of death or serious injury is very low.[16] No round is 100 percent safe under all conditions. The user must determine if use of such force is justified and compare the possible outcomes with the alternatives available. Certainly, the probability of serious injury from any non-lethal/less-than-lethal round is far lower than a bullet from a 9-mm pistol, but I decided to find out for myself.

The Los Angeles skies were overcast on 16 March 1998, when Lieutenant Sid Heal and I met to conduct tests on the Sheriff's Office firing range. Accompanying us was professional photographer Robert Knight. The television weather forecaster had predicted clear weather and warmer temperatures. Sid and I had decided that since no penetration test results were publicly available that could counter the emotional claims made against low-impact projectiles, we could run some quick ballistics tests. While the lighting was not ideal, it was good enough to capture the results.

The tests were designed specifically to draw a comparison between various non-lethal munitions and the police officers usual alternative, the 9-mm pistol. I felt it was time to conduct some side-by-side examples that could unambiguously answer the charges of opponents of non-lethal weapons that they caused too much physical damage. My counter, "Compared to what?", was about to be answered.

The tests were conducted with both gelatin blocks developed for ballistics tests, and for dramatic effect, watermelons. A gelatin substance is used as it replicates the hydrostatic shock deposed by bullets on a human body. The melons were chosen as they had been used in a television special to portray non-lethal weapons as causing excessive injury and pain. During that program it was shown that a bean-bag round would puncture a watermelon. What they failed to show is the damage caused by a traditional bullet. The first round Sid fired was a Defense Technology 21 BR bean-bag cartridge. The 12-gauge shotgun round contains a small amount of metal shot wrapped in a canvass bag. Fired from a distance of twenty-one feet, it penetrated the upper right quadrant of the gelatin to a depth of five centimeters or about two inches. It should be noted that the calibration for this gelatin consists of a BB traveling at a velocity of 590 feet per second, penetrating the block to approximately eight centimeters when its temperature is at four degrees Celsius.

An excellent marksman, and long-time SWAT team member, Lieutenant Heal then fired his standard 9-mm pistol loaded with Ranger SXT ammunition into the center of the same gelatin block. The round penetrated thirty-six centimeters or about fourteen and a half inches. Further, as can be seen clearly in the photographs, the round was spinning as it passed through the gelatin. That translates to extensive tissue damage when such a round strikes a human body.

Additional non-lethal munitions, including stingballs, wooden batons, and foam rounds, demonstrated they could either penetrate the gelatin to a slighter degree than the bean-bag rounds, or would bounce off but produce shock waves in the target. Pictures of the results are in the photograph section.

Most dramatic was the watermelon demonstration. As can been seen in the photo insert, the foam batons either bounced off the melon, or barely stuck in the surface. The bean-bag round did penetrate the melon and cause some fracturing although it remained whole. The round did not exit the melon. However, the impact of the standard 9-mm round caused the watermelon to disintegrate, scattering the pieces of a wide area.

As has been acknowledged, people have been seriously injured with low-kinetic impact munitions. In a few cases, deaths have resulted. But, if it is assumed that use of force was necessary and authorized, those injuries should be fairly compared with the damage caused by a standard lethal bullet.

Low-KE rounds offer law enforcement officers and troops in peace support operations attractive options between shooting and not shooting. Unlike irritant sprays and foam technology, these rounds offer adequate standoff distance. A problem with many non-lethal systems is reach. The question to be answered, specifically, is, Can it provide more range than a stone? Some of the extended-range, low-KE munitions provide that safe distance. Most of the low-KE rounds can be fired from existing weapons, which is a distinct advantage for police officers and peacekeepers. The rounds, while non-lethal, do cause some blunt trauma. Next we will examine an emerging category of weapons that may overcome even that concern.

9

ACOUSTICS

With a great shout, the walls fell down flat.
—*Joshua 6:20*

The best known use of acoustic weapons dates to the time of Joshua and the battle for the seemingly impenetrable walled city of Jericho. To break the siege, Joshua was commanded to have seven priests, each bearing a ram's horn, circle the city on six consecutive days. Each time they traveled around the walls, they were to blow loudly on their trumpets, but the accompanying soldiers were to remain silent. On the seventh day, the priests circled the city seven times, continuously making noise from the rams' horns. Upon completion of the seventh circumnavigation, the soldiers of Israel let forth a great shout. At that point, the Bible states, the heavy stone wall collapsed, and Joshua's men entered Jericho and killed the inhabitants.[1]

The cause of the structural failure of the stone walls has long been debated. The commonly accepted version infers that divine intervention led to the collapse. Another interpretation suggests that the acoustic vibrations set up by the rams' horns actually weakened the internal structure of the walls to the point that they became incapable of supporting their own weight. As we shall discuss, there may be some scientific support for that claim. Whichever is correct, the incident remains one of the most well-known tales of Biblical warfare.

A more recent, and very crude, attempt at acoustic warfare took place during Just Cause, the U.S. invasion of Panama. One of the primary missions was to capture Manuel Noriega and return him to face trial in the United States on drug charges. During the early days of the invasion, with help from his many loyal supporters, Noriega managed to elude capture. As the U.S. forces tightened the noose on him, Noriega surreptitiously slipped into the papal Nuncio of the Catholic Church and requested asylum. Honoring the age-old tradition of religious sanctuary, our soldiers were ordered to stay out of the grounds. Arresting Noriega under such conditions would set extremely poor precedence, not to mention making very bad press.

General Maxwell Thurman, the newly appointed Commander in Chief of U.S. Southern Command, ordered another tactic employed. He had high-powered loudspeakers set up outside the Nuncio and directed that heavy metal rock music blare around the clock. The attempt was insufficient to drive Noriega out, but it did anger many of the adjacent residents. However, the troops generally enjoyed the music, thus demonstrating the cultural gap between General Thurman and the young soldiers.[2] While the music did receive a lot of press at the time, these actions are probably more in line with a poorly designed psychological operations project than a true acoustic weapon.

An important adjunct in non-lethal weapons development, acoustic technology has made some major advances in the past few years. Both matériel, as in the walls of Jericho, and personnel, such as Noriega, can be attacked by acoustic weapons. In general, these capabilities have been overlooked. However, at the end of this chapter, a recent breakthrough that could revolutionize the importance of acoustic weapons will be revealed.

There are many military law enforcement applications for acoustic weapons. In civil disturbances and peace support operations, there is a need to gain control of violent situations with minimal force. These weapons can be used to drive people away from a selected area or to enforce a safety zone between troops or police and potential attackers. By keeping people at a distance greater than they can throw rocks or other missiles, they provide a margin of safety currently not offered by other non-lethal systems. Unlike chemical agents, an acoustic field does not contaminate the area; thus, no cleanup is necessary. An incremental application can be envisioned. Tunable systems can be employed initially at low levels. Should violators not respond, intensity can be increased until compliance, voluntary or otherwise, is obtained. This can be accomplished either by increasing power or by moving the source closer to the target. Another advantage is that acoustic waves are efficiently transmitted through smoke, fog, and dust. That is not true for electromagnetic weapons.

Acoustic beams might be used as point-defense tools. Consider a situation similar to the takeover of the U.S. embassy in Tehran in 1979. Embassies are outposts, and external security is the responsibility of the host nation. When that fails, the guard force on duty is usually insufficient to hold off an adversary, except for a brief period of time. However, that may be a very important period during which cipher machines and other sensitive equipment are destroyed and, if possible, key people evacuated. It is understood that without external support, the physical compound will be lost. Therefore, lethal force, while possibly authorized, is very unwise. The guards and other embassy personnel will undoubtedly become prisoners. The crowd is already hostile, and killing some of them as they break in is likely to lead to exceptionally harsh treatment. Acoustic weapons offer an ideal non-lethal alternative for this scenario. With international trade expanding, businesses may be faced with similar situations.

The objective of antipersonnel acoustic weapons is not to create untenable sound, but, rather, to vibrate the targeted people physically. In order to establish barriers or to cause people to move from a restricted area, it is necessary to do more than make loud noises. The countermeasure for that is simply to cover one's ears.

By way of comparison, the effects of acoustic weapons are not unlike long-term exposure to rock music. Some rock concerts are so loud, often exceeding 110 decibels (dB) that pronounced hearing loss has occurred in people of relatively young age who attend frequently or participate in a band. Most rock concerts may peak for very brief periods of time—measured in milliseconds— up to 130 dB. Note that aircraft engines at a distance of about 100 meters will register only 100 dB. Acoustic weapons will range from about 120 dB to as high as 170 dB. At levels above 150 dB, internal injury can occur.

There are three levels of acoustic frequencies that might be applied as weapons: infrasound, audible sound, and ultrasound. *Infrasound* is at the low end of the spectrum. Some believe there is little practical use for infrasound but in the natural environment, it can cause illness and even damage to buildings.[3] Exposure to low-energy infrasound for long periods has been listed as the attributable cause for "sick buildings" or "sick cities." Due to the low frequency, it is very difficult to screen out and can easily penetrate inside buildings. Higher intensities can cause nausea and disorientation.

Audible sound, in frequencies from 20 to 20,000 hertz, can be applied to influence behavior, as most people are sensitive to very loud noises. Further, sound plays a major role in our psychological makeup and behavior. The entertainment industry figured that out long ago. Movies use music and other sounds to manipulate the audience. Remember the heartbeat sound used so effectively in the landmark movie *Jaws?* It was not accidental that the sound used when the great white shark was about to strike would evoke a visceral response in those watching. The impact of the sound alone was sufficient to keep many people out of the water that summer. Similarly, there is a reason that fingernails scraping across a blackboard will send shivers through almost everyone. Even thinking about it now might elicit that same response. Through physical and psychological maneuvering, there is a role for audible sound in non-lethal weapons.

Ultrasound, frequencies above 20 kilohertz, are well known to the medical community. It was first used by an Austrian psychiatrist, Karl Dussik, in 1942 to locate tumors in the brain visually. Since then, ultrasound has been used in both diagnostic and healing procedures. Although no hazard has been identified with diagnostic ultrasound, people exposed to high dosages report noticeable heating effects that could lead to injury.[4] However, the high frequencies involved do not propagate as well as the lower frequencies and can easily be externally blocked. I am not aware of any attempt to develop weapons in the ultrasonic range.

Given the advantages of better propagation, most work on acoustic weapons has been done at the lower end of the spectrum. At low frequencies, it is possible to cause internal vibrations that generate a number of effects, depending on the frequency and power levels employed. The effects cannot be overcome through hearing protection, personal perseverance, or being impervious to pain. Of course, care must be taken in the use of low-frequency sound so as to prevent permanent injury, or in extreme cases, death. There are persistent rumors about a French researcher, Professor Vladimir Gavreau of Marseilles, who was developing high-power infrasound weapons. Gavreau began his work after an unanticipated exposure to infrasound from a defective ventilator. When he discovered a method to direct acoustic energy, he turned to weapons development. He reported that his tests, which caused internal vibrations of internal body organs, brought "unanimous and vociferous protests from members of nearby laboratories." The available information suggests that a colleague, Dr. Levavasseur, accidentally was subjected to a blast from the weapon and died as his internal organs turned to jelly.[5] However, no matter how persistent the rumors, others who have researched the field have failed to confirm this story. It is recanted as one of the enduring myths of acoustic weapons. But acoustic weapons have been explored for a long time.

The first developmental obstacle that had to be overcome was size. Early attempts at introducing acoustic weaponry were limited by the enormous size requirement for speakers. Those dimensions were determined by the wavelength of the low-frequency sounds desired. The physics problem was that, simply, the lower the frequency, the larger the speaker had to be. The low frequencies were needed to produce the physical effects.

During World War II, German and Austrian scientists were employed in the development of powerful acoustic weapons. Reichsminister Speer became convinced that an acoustic-beam weapon could be developed that could incapacitate troops at ranges up to twenty kilometers. Hochtal Laboratory at Lofer, Austria, was assigned the research and development task. The director, Dr. Wallanschek, formerly of Telefunken, believed the range objective was absurd. However, he was able to produce intense sound that could incapacitate people at distances of sixty meters. His technique, controlled intermittent combustion, was the forerunner of today's technology. They probably were able to produce in the order of 100 kilowatts of power. A military utility problem was that the reflectors for the sound generators were greater than ten feet in diameter.[6] There is some evidence that the Germans attempted to develop acoustic antiaircraft weapons. Conventional antiaircraft weapons could not reliably strike bombers flying above 15,000 feet, and as saturation bombing of German industry was increased dramatically, new weapons were needed to hit the high-flying aircraft. Any and all possibilities were considered. One such weapon involved the detonation of gases that produced pulses designed to break off the wings of bombers. It was too late in the war, however, to be fully implemented. Present-day aircraft fly far too high to be at risk from acoustic weapons.

However, following World War II, others, including some American inventors, did follow up on the German experiments. Acoustics was one of many areas of science explored by Guy Obolensky. An innovative and entrepreneurial researcher, Obolensky explored several unique areas from which others shied away. Learning of the prior German work, he conducted several experiments demonstrating that acoustic weapons could project sufficient force to make a viable weapon. The work led to the fundamental patent in the field of physical and chemical effects from sound waves, which is held by another inventor, Robert P. Shaw.[7]

Obolensky noted an interesting physiological effect that could be induced by high-power infrasound. He found that, frequently, exposure to critical frequency would cause the bowels to release involuntarily. The concept was considered interesting as a riot control measure. A formal proposal was submitted to the Navy with the title of SuperPooper. The idea was reportedly rejected by a Marine general who felt the potential weapon to be too undignified.[8]

The current leader in development of acoustic weapons in the United States is Scientific Applications and Research Associates (SARA), a small company located in Huntington Beach, California. The president, Parvis Parmani, enticed several innovative scientists, including Tim Rynne and John Dering, away from a major aerospace company. Each was searching for a challenging work environment, one in which they could explore emerging technologies in ways not appreciated by corporate giants.

Among Dering's past excursions was a study of ancient applications of acoustics. During his studies, Dering located the work done by the Germans and others, years before. Based on this knowledge, he convinced his colleagues that acoustic weapons could play an important role in the emerging nonlethal weapons field. Jointly, they developed a proposal that was subsequently funded by the Army's ARDEC and ARPA to research and develop prototype acoustic weapons. By 1993, SARA conducted the proof-of-principle demonstrations of an infrasound weapon system.[9]

Similar to the German approach, SARA employed repetitive detonation to create intense toroidal vortices. The fuel is a simple mixture of methane and oxygen. This combustion-driven acoustic source emits pressure waves at greater than 130 dB, sufficient to incapacitate anyone within a targeted area.[10] In fact, most attempts at developing intense sound devices have used the repetitive-explosion technique.

At Los Alamos, Ricky Faehl and George Nichols also developed a prototype acoustic system using propane. In their experiments, they were able to add another innovation that is important for making compact acoustic sources. Using multiple tubes, they cut small holes in each tube so that the sound resonated between them. This phased-array system allowed devices to be reduced in size at least an order of magnitude.[11]

In Europe, Bengt Wigbrant, of the Swedish National Defence Research Establishment, was the project manager for the development of their Vortex-Generator.[12] Their system, which also has been demonstrated, was fueled by propane that was detonated in a combustion chamber. A tube placed in front of the chamber formed the pulses of energy into a whirl that was projected toward the target. It could easily be thought of in terms of invisible smoke rings projected one after another. The gas whirls are energy packages that move through the air at speeds of forty to sixty meters per second and pack a significant wallop. The name applied to this system was *High-Energy Whirls,* or *HEW*. In August 1997, SARA, applying the same principles, was able to generate ring vortices two feet in diameter that traveled more than the length of a football field at 70 meters per second.[13]

During development of the Swedish system, they noted the early German experimentation and urged that caution be taken when testing the elementary devices. According to their sources, a 1945-era generator had been tested against some large pine trees. The records indicated that when the whirls hit the top of the tree, "limbs as thick as an arm were cracked and fell to the ground."[14] A friend of mine in the German Air Force observed a demonstration and confided in me his concern that the current Swedish system might cause more physical damage to humans than would be acceptable. However, that is an engineering problem that can be overcome.

Before the end of the Cold War, the Soviet Union experimented with acoustic weapons. U.S. intelligence reports indicated that Soviet scientists had been experimenting with a wide range of sounds in efforts to determine the physiological and psychological effects. In the infrasound zone, they discovered that exposure for extended periods of time brought on impairment in tracking ability, choice-reaction time, and peripheral vision. At seven hertz they reported difficulty in mental activities and precision work. High-intensity infrasound induced sensations of panic in some subjects.

When they tested the audible frequencies, they found that certain sounds could also disrupt thinking and frequently produced drowsiness. At times, with prolonged exposure, the effects were strong enough to have test subjects falling asleep on their feet. In addition, ultrasound was reported to create fatigue and general weakness.[15]

The Soviets then took this fundamental research and converted it to weapons design. There were later reports that they had an external acoustic system that was mounted on some tanks. The system was designed to keep dismounted infantry soldiers from approaching from blind spots and climbing up on the tanks. In cities with limited maneuvering space and when operating against civilians, the threat from primitive weapons can be considerable. The *Molotov cocktail,* a simple bottle filled with gasoline and capped with a rag fuse, is now known around the world as a poor-man's antitank weapon. Acoustic devices could keep people with primitive devices far enough away to prevent them from striking the tank with such a weapon.

Both the Soviet Union and Sweden have also experimented with beam convergence. The process involves generating two different frequencies and pointing them toward the target. It is at the point of convergence that the effects are noticed. In the acoustic range, it allows for line-of-sight, covert communication without the hindrance of any electronic device. This would be advantageous, for example, to an agent buying drugs. He or she could undergo a thorough inspection for "wires" and found to be "clean." Although they could not transmit, the agent could be warned of danger or when the bust was about to go down. From a weapons perspective, the psychophysiological effects would only be present at that precise location. It was reported that the Soviets tested this approach as a means for clearing buildings or underground structures. The concept entailed creation of synchronous, rhythmic vibrations approximating an earthquake. It was believed that there is a natural aversion to those frequencies, causing continued habitation to be untenable.

Developing directionality for acoustic weapons is very important. One problem with early attempts was that everyone in the area could be adversely affected by the weapons. In fact, some of the infrasound effects were discovered through accidental exposure. These omnidirectional adverse effects limited the potential usefulness of the early acoustic systems. In the United States, SARA has demonstrated advanced beam steering techniques that are necessary for weaponizing acoustics. Based on the concept of multi-element RF antenna engineering, SARA developed an end-fire system that provides the user with a wide variety of acoustic capabilities.[16] They can tune the antenna from narrow to wide angles, depending upon the size and location of the threat.

Additionally, SARA engineers have adapted several of their acoustic systems for mounting on aircraft. These are designed for crowd control and area-denial missions in which other technologies are not available or sufficient. For airborne applications, it is imperative that the acoustic output can be carefully controlled.

While most acoustic work has focused on antipersonnel applications, some work has been done on antimatériel weapons. The mythology of acoustic interactions with materials is not limited to the battle for Jericho. Ancient Tibetan applications of sound for levitation of heavy objects have been documented in recent years. A famed Swedish aircraft designer, Henry Kjellson, observed and recorded heavy stones, each about a 1.5-meter cube, being lifted to a position 400 meters above them by monks using musical instruments.[17] The technique required thirteen drums and six trumpets, complemented by about 200 priests, all placed in a specific pattern. When the trumpets were blown at about two blasts per minute, the stones reportedly would rise and follow a prescribed trajectory to the top of the cliff. Other, poorly documented, stories have suggested that a similar procedure may have been used to move stones at the pyramids in Egypt. On a more scientific note, recent advances in ultrasound have been used to levitate lightweight items close to the source.

In addition to levitation of objects, it is known that materials can be destabilized with acoustic energy. Other indications of the interaction of sound on stone objects include complaints that have been lodged by the U.S. Parks and Recreation Service about damage to the bridges of red rock at Arches National Monument in Utah. The damage has been caused by U.S. Air Force fighters that frequently overfly the area on combat maneuvers.

Dean Barker, an Old Crow from the World War II era, has continued to explore military applications of acoustic technology. He stated that, in experiments he conducted for the Army twenty years ago, they were able to move concrete walls a slight distance. Barker also reported that they had successfully destabilized critical metal elements. He claimed to have an acoustic device that could be placed next to railroad tracks that would cause them to weaken. The structural fatigue was not visible to the naked eye, but the track would disintegrate when a train ran over it.[18] Clearly, there is a military mission for acoustic technology that can prevent enemy material from being used effectively.

One area that was explored by Soviet researchers but received little attention in the West was the use of acoustic systems in conjunction with chemicals to enhance their effects. While some of the reported effects were intentionally fatal by initiation of anaphylactic shock in test animals, non-lethal approaches could also be considered. As an example it may be feasible to apply subcritical doses of a substance to one or more people, then later induce hypersensitivity with an infrasound device.[19] While this technique would surely come under extensive criticism, its application by those not constrained by international treaties makes the possibility worth exploring from a defensive posture.

Probably the most important breakthrough in acoustic weapons was accomplished by the engineers at SARA, with the support of Richard Dickhaut, until recently the president of Spectra. Dickhaut has extensive experience in studies employing electroneurophysiology. Specifically, he had designed a mechanism for using an electroencephalograph (EEG) to monitor the brain's processing of specific thoughts. Using that knowledge, he could then determine when a designated thought had been processed. Working with SARA, Dickhaut and his colleagues were able to design a system that can create neuropsychological distress in the central nervous system through use of modulated pulse. The technique is called Pulsed Periodic Stimuli (PPS). The technique can be applied in situations where it is desirable to cause perceptual disorientation in targeted individuals. This is important, as it is the first acoustic weapon that does not rely on high intensity to cause the desired effects. Rather, low-intensity, pulsed, acoustic energy can induce fairly strong effects in humans.[20]

10

INFORMATION WARFARE

Information warfare, cyberterrorism has, like the six-gun, been called the "great equalizer."
—*Senator John Kerry,*
The New War

At 3:23 P.M. on Friday the stock and bond markets, driven into a free-fall by offshore trading, were automatically closed when the Dow plunged through the daily maximum allowable 500-point drop level. Something worse had already occurred, but it was not yet recognized. At 12:00 noon the Wall Street computers all crashed—simultaneously. Due to lag time between input of transactions and output of records, the extent of the damage had gone unnoticed. However, all trades that had taken place in the market panic between noon and closing—and there were millions of them—failed to be recorded. The financial world was in chaos—except for those who had precipitated the event.[1]

Thus runs one theme in Tom Clancy's 1994 novel *Debt of Honor*. It is a poignant example of an information warfare (IW) attack on the financial infrastructure of the United States. In the book, a quick economic recovery was made, followed by a sound military victory. Clancy, recognized for his patriotism, always has America emerge victorious. In this case, I believe he was a bit optimistic. A concerted IW attack could be devastating, and a quick recovery is not assured.[2]

Snippets of IW have permeated our culture through the entertainment industry. Attacks, albeit sometimes diversionary, have been employed by many villains. In *Die Hard II,* the bad guys electronically took over Washington, D.C.'s air traffic control and caused planes to crash as a means of extortion. Meanwhile, star Bruce Willis waited to locate his adversary so he could respond with brute physical force. *Hackers* depicted skilled teenaged computer freaks on Rollerblades tangling with professionals of organized crime. In *The Net,* Sandra Bullock had her identity erased when she discovered an illegal operation to sell extremely sensitive computer security codes. Robert Redford, leader of a highly sophisticated IW team, pursued a mythical "universal decryption code" in *Sneakers.* As early as 1983, *Wargames* showed computer-literate youngsters breaking into NORAD computers and nearly initiating a nuclear strike against the Soviet Union. Then a novel idea, movie reviewer Roger Ebert called it one of the best films of the year.

Information technology plays a pivotal role in non-lethal warfare. When the basic concepts were being formulated, information technology was incorporated as a subset of non-lethal weaponry. However, in the intervening years, IW has taken on a role of such magnitude that it deserves independent status. Under various names, including *command and control warfare, computer warfare, cyberwar,* and some aspects of *electronic warfare,* IW concepts and funding have escalated dramatically. Still, there is a need to address the fundamental aspects as they relate to non-lethal alternatives. The direct impact of an IW attack is non-lethal, but secondary effects may not be.

In addition to fictional accounts of IW attacks, the media have reported numerous real ones. Too frequently Department of Defense computers have been targeted—sometimes by joyriding young hackers; sometimes by foreign operatives. Several studies have evaluated the vulnerability of DOD networks. All agree that many attacks have occurred and will continue to occur. In one report, the Defense Information Systems Agency (DISA) estimated that in 1995, approximately 250,000 unauthorized probes were made of the unclassified computer systems of the Defense Department. They went on to estimate that about 160,000 were successful.[3] Other studies estimated that as many as 90 percent of such illegal entries went undetected. Break-ins of classified systems have been attempted, but no successful breaches have been openly reported.

The scope of the IW problem staggers the imagination. The United States is the most information-intensive society in the world—and therefore the most vulnerable. However, all of the developed world, and much of the developing world, is rapidly becoming information dependent. Only self-sufficient agrarian economies are relatively secure from IW attacks. Information technologies offer a paradox. The spread of information is a strength, while openness of the systems constitute weaknesses. Therefore, democracies, with their innate openness, are inherently vulnerable to IW attacks.[4]

The objectives of IW are very broad. In one of the first major books on the topic, *Information Warfare: Chaos on the Electronic Superhighway,* Winn Schwartau discussed the breadth of reason to engage in such conflict. He stated that IW is about money, power, fear, arrogance, politics, defiance, disenfranchisement, control, and, ultimately, survival.[5] To put a dollar figure on current U.S. losses, Schwartau estimated that $100 billion to $300 billion per year "slithers through the Global Network and out of control." Of course, long before Schwartau's book, Alvin Toffler and his wife Heidi had predicted the impact of information technology—for which they coined the well-known term *The Third Wave.* Their book, *War and Anti-War,* detailed the influence of digitization and information on the modern battlefield, and suggested what would come in future conflicts.[6]

Defining information warfare is a formidable task. Many of the key IW documents dodge the issue by talking about it without using a definition. A RAND study suggested that IW is evolving rapidly in response to the "information revolution" and therefore definitions are at best imprecise. In fact, the Defense Science Board IW-D study observed, "We lack a common vocabulary."[7] Air Force Chief of Staff General Fogelman called IW "the fifth dimension of warfare."[8] Martin Libicki, a professor at the National Defense University, lamented that some definitions pertaining to IW were so broad as to "cover all human endeavor."[9] To provide some context, Dennis Richburg, Technical Director and IW expert for the U.S. Air Force Air Intelligence Agency at Kelly Air Force Base in San Antonio, Texas defined Information Warfare as "Actions taken to preserve the integrity of one's own information systems from exploitation, corruption, or destruction while at the same time exploiting, corrupting, or destroying an adversary's information systems."[10] Note that the definition encompasses all information systems, *not* exclusively military systems.

The definition also addresses both defensive and offensive capabilities. For several years, those working in the field of IW were proscribed from mentioning offensive actions. Based either on naïveté or, more likely, on turf struggles, those responsible for classification kept information on offensive measures suppressed, even in the developmental community. It was almost as if they believed no one would assume that the United States would develop and employ offensive IW weapons. While these overly restrictive classification actions did protect weapons operating parameters, a reasonable venture, they also inhibited participation by many skilled, innovative—*and loyal*—computer scientists (another example of *cerebralcentrism*). The efforts successfully blocked the infusion of creative ideas for new IW techniques. In addition, these restrictions tended to keep U.S. forces in the dark about the technical threats they could be facing. Of course, beyond the military comes the thorny issue of what knowledge of IW should be passed to the civilian sector in order that businesses and individuals might be warned about threats to their information systems.

Recent articles have begun to contemplate the strategic importance of the issues raised by information warfare.[11] Several scholars note the potential for a surprise attack similar to that posited by Tom Clancy and liken it to the infamous sneak attack on Pearl Harbor.[12] The Defense Science Board has been waving a warning flag for several years. Their first recommendation is to designate an IW focal point.[13] While absolutely necessary, this action only solves a piece of the problem since few understand that IW is not exclusively a military issue. This can be problematic when examined from the traditional "roles and missions" perspective taken by most government institutions. Warfare is normally the domain of the Defense Department. Information permeates every aspect of society. The interdependency of information systems is outlined in the 1997 Department of Defense joint pamphlet titled *Information Warfare.* It addresses the Global Information Infrastructure (GII), in which is embedded the U.S. National Information Infrastructure (NII), which in turn encompasses the Defense Information Infrastructure (DII).[14] The DII is not a separate, secure, or encapsulated system, but shares common communications systems with civilian networks. The relationship is seamless, using both terrestrial and satellite communications networks that span international boundaries. While a few DOD-unique systems exist, the vast majority of Defense communications travel via shared systems. Likewise, databases are also increasingly shared with civilian users.

The joint pamphlet on IW suggests that DOD, other government agencies, academia, and industry are partners in IW. They note that a team approach is essential for development of a comprehensive strategy—and it is.[15] However, simple coordination between participating agencies and civilian elements will not be sufficient to defend the information infrastructure and prosecute IW. While embryonic efforts to address these intricate and perplexing problems are being initiated and coordinated at the highest levels, the issues are so complex that they will require extensive mental and financial resources to solve. It has been recommended that the focal point for IW should be located no lower than the Executive Office of the President.[16] A step in that direction has been taken with the establishment of the President's Commission on Critical Infrastructure Protection.

Information warfare transcends traditional conflict. In today's business world, which is very dependent on information, IW has special significance when engaged in economic warfare. Martin Libicki notes that "the marriage of information warfare and economic warfare can take two forms: information blockade and information imperialism."[17] The information blockade infers the ability to restrict the flow of data to the threatening nation. While ground-based communications can be cut and the jamming of some signals is possible, stopping all transmissions, especially from direct-broadcast satellites, would be extremely difficult. Nonetheless, in a world that thrives on real-time information, any restriction could have serious consequences. Of course, this problem is complicated if the adversary does not have a fixed geographic location.

The information blockade could be accomplished via control of financial systems and blocking of information necessary for senior officials on which to base decisions. Similarly, incorrect information could be inserted into the targeted country. After a few bad decisions were generated, officials would distrust all information that could not be independently verified. This would quickly lead to economic chaos.

Counterterror experts Robert Kupperman and Frank Cilluffo, writing in *The Brown Journal of World Affairs,* compare the problems of weapons of mass destruction (WMD)—nuclear, chemical, and biological—with information warfare. They suggest that IW offers another form of WMD: *weapons of mass disruption.*[18] Certainly, IW will become one of the favorite weapons of modern terrorists. The costs of entering the IW game are very low, and the potential payoffs very high. The nature of the game also affords a high degree of physical security to attackers. They may sit with their computers and modems anywhere in the world, patch in through complex telecommunications routes, and strike at will. Further, if properly executed, it may be difficult for the victim to distinguish an actual attack from an unexpected coincidence. Unlike conventional military force, it only takes a few skilled individuals to cause extensive damage. Due to economic conditions in many parts of the world, there are many computer-talented people ready to sell their skills to the highest bidder. Since IW equipment costs are in thousands of dollars, not millions, as for conventional weapons systems, the highest cost is probably paying personnel.

From a criminal perspective, when the cost of doing business is balanced against the potential gains, the outcome is inevitable. Companies involved in electronic transfer of funds move enormous sums of money. As an example, IBM's Swift transfer system reportedly handles $3 trillion daily. Diverting thousands of dollars, or even millions, would be trivial by comparison. Even programs that simply round off a few cents on each transaction can illegally collect large sums of money. At the end of each business day in America, substantial amounts of money cannot be accounted for. In fact, the daily amount of "missing" dollars in the U.S. Treasury is so great that the figure is kept from the public.

Another disquieting factor is that the skill level required to engage in IW is declining. Initially, only a very few, highly educated computer scientists held the requisite skills to create viruses and surreptitiously break into information systems. One budding young American genius, in coordination with several offshore experts, led computer spy hunters on a two-year chase through an electronic labyrinth. His capture, by astronomer-computer scientist Cliff Stoll of Lawrence Berkeley National Laboratory, was an embarrassment to many. The American involved, who had written the key programs—including one that froze 2,000 computers—turned out to be Robert T. Morris, a graduate student at Cornell University. The trail led to the top national code breakers. Robert Morris was the son of the National Security Agency's chief scientist of their supersecret computer security center.[19]

With the support of user-friendly hardware and software, accompanied by step-by-step instructions, even those with modest skills can attempt to enter the game. Companies now offer mix-and-match virus programs in which the operator need only move the cursor and click to design the virus of his or her choice. Books such as *Secrets of a Super Hacker* provide instructions to anyone with twenty dollars.[20] If you don't want to buy a book, you can ask other hackers for advice. With names like Darkcyde, the Dark Wizard, Dark Angel, Dark Phiber (hackers like dark names), the Legion of Doom, PsychoHackers, Bad Karma, and even Lucifer, they are eagerly waiting to help. There are catalogs of hackers' products, one of which calls itself the "world's most dangerous catalogue"—and it may well be. Their product line includes 4,000 viruses—or you can buy the virus lab and create your own. They advertise instructions on phreaking, stealing cell phone identification numbers, and how to build a credit card writer. On the more physical side, they actually have electronic warfare weapons, including jammers, EMP weapons, and the like. In case these won't get the job done, instructions for building your own nuclear weapon are listed. With such readily available help, the probability of both mischief and mayhem on the Internet is rife.

Not all IW takes place on a global scale. In fact, low-level IW plays a role in our daily lives. The most notable examples are hackers. Capturing the fancy of many, the Discovery Channel even has a Hackers' Hall of Fame on their Web site. Usually independent souls functioning behind a veil of anonymity, hackers nefariously break into information systems and commit crimes that run from mildly amusing to major theft—and worse. For hackers, nothing is sacred. Take, for instance, the dastardly intrusion that happened immediately following the 1997 Memorial Day release of Steven Spielberg's megahit *The Lost World: Jurassic Park.* As with many movies, Universal Studios established a Web site where interested people could obtain information about the motion picture and, more importantly, find pictures of dinosaurs. Cleverly, hackers entered the site and replaced the fearsome image of *Tyrannosaurus rex* with a motley version of a prehistoric fowl. The new caption read, THE LOST POND: JURASSIC DUCK!

Then there were the ventures of Dark Dante, the moniker of a young hacker by the name of Kevin Poulsen. An enterprising sort, Poulsen learned that a Los Angeles radio station was going to give away a new car. It would go to the one-hundred-and-second caller at a specified time. The radio station, KLLS-FM, expected the giveaway to be like a lottery, with the prize going to the lucky caller. However, Poulsen had other plans. As Dark Dante he surreptitiously took over all the local phone lines. No one else could call in. And it was not just any car, it was a Porsche 944 that Poulsen won as the hundred-and-second caller. Of course, he was also caller 1 to 101 as well.

Another innovation came from Cap'n Crunch, aka John Draper. His handle was apropos. From a whistle found in a cereal box, Cap'n Crunch learned how to get free long-distance phone calls by making precise sounds that represent numbers. Draper is credited with developing a special brand of hacking called *phreaking.* Of course, this later evolved into the art of stealing cell phone identification numbers and the loss of millions of dollars in fraudulent phone calls. The people using these illicit techniques became known as *phrackers.*

On a larger scale of crime, Vladimir Levin was able to get Citibank to provide his bank accounts with $10 million. Levin was a graduate student operating not in the United States but from St. Petersburg Tekhnologichesky University in Russia. He entered the banking system via multiple foreign telephone connections and arranged the transactions. His success came to an end when he was arrested in London's Heathrow Airport in 1995.

One hacker has the distinction of making the FBI's Most Wanted list: the infamous Kevin Mitnick. Described as an antisocial computer wizard, he had a long history of hacking violations and has done millions of dollars' worth of damage. In 1988, Mitnick was arrested for breaking into Digital Equipment

Company's computer network. They reported that he caused $4 million in damage and stole $1 million worth of software. For those crimes he served eighteen months in prison and at a treatment center. After his release, Mitnick went right back to hacking. His criminal acts included breaking into thousands of computer programs and the theft of 20,000 credit card numbers all over the United States. Pursued for two years, he evaded authorities at every turn. Mitnick's undoing came when, using highly sophisticated maneuvers, he reportedly hacked his way into another computer expert's home system on Christmas Day, 1994. Tsutomu Shinomura, himself a security expert, decided to track down the intruder. Less than two months later, Mitnick was in custody charged with computer fraud, wire fraud, cellular phone fraud, and illegal use of phone devices. Despite making the Most Wanted list, Mitnick plea-bargained federal prosecutors down to a very light sentence for someone on his third conviction: eight months' jail time![21] No wonder people don't take white-collar crime seriously.

A secondary cost generated by hackers is the sale of computer security systems. It is predicted that anti-hacker software sales will increase from $1.1 billion in 1995 to greater than $16 billion in 2000.[22] The President's Commission on Critical Infrastructure Protection noted the increased likelihood of computer terrorism and recommended that research and development alone, now $250 million annually, should be increased $100 million per year until $1 billion is provided on a yearly basis.[23]

While hacker attacks against government computers occur on a continuous basis, a February 1998 series of attempted break-ins caused Deputy Secretary of Defense John Hamre to call them "the most organized and systematic attack the Pentagon has seen to date." Between 14 and 25 February, 800 networks of the Army, Navy, Air Force, and other agencies were intruded upon. The culprits were all teenagers. One calling himself the Analyzer lived in Israel, while his two accomplices attended Cloverdale High School in northern California. When asked about motive, one hacker, who identified himself as "Makaveli," stated to hacker journal *AntiOnline* that he hacked for "power, dude, you know, power." This is another example of a burgeoning and pervasive problem.

The problem has finally become recognized for its international level of importance. In December 1997 the justice and interior ministers of the United States, Russia, Germany, Japan, Canada, Italy, France, and the United Kingdom agreed to coordinate efforts in fighting information crimes.

A new IW area of concern for U.S. forces will be the integrity of their equipment. In wars past, our troops usually had confidence in their war-fighting matériel. While a few isolated cases of fraud were noted and some poor design work was done, the equipment was trusted to be sound. Normally, it was American made. Specialized electronics were developed in military laboratories or under the guidance of a few trusted contractors. However, times have changed. The military is almost totally reliant on commercial products that have been adapted to military use. More and more complex weapons systems have parts that are made by a multitude of subcontractors. Remember, for rockets to fire and accurately hit their target requires the 100 percent reliability of many subcomponents. A small deviance at any level could cause a failure of the missile to hit its target. The concern is that if a stealth virus could be implanted during the manufacturing process and then activated externally, possibly by an electromagnetic signal, our high-tech missiles might go awry. This is only one of many possible scenarios for corruption of weapons systems. The positive side of such technology is to prevent use of U.S. weapons systems against us. The issue came to the fore when we were supplying Stinger missiles to the anti-Soviet mujahideen rebels in Afghanistan. There was reason to worry; many of those missiles remain unaccounted for today.

From a defensive perspective, there is only a limited amount of hardening that can be done, and the cost of hardening all military systems is prohibitive. Redundancy and proper use of basic security measures will go a long way toward protecting our operational capability. To defend the core of the NII and MII, it has been recommended that a process called "minimum essential information infrastructure" be explored.[24] This process would determine the key assets that must be protected at all costs. Additionally, our forces must be capable of operating from a position of *digital default.*[25] This means they must be able to function effectively without relying on their digitized equipment. However, it is the digitization that has provided the United States with the high-tech capability to fight outnumbered and win. Without our advanced systems, the odds for winning decrease dramatically.

The concept of digital default is wise and prudent. The methods by which computers can be attacked are far more varied than most people know. Everyone who uses a computer is aware of the danger of viruses that can wipe out their programs. They understand that they should not download executable code from unknown sources. But that is only a small part of the problem. As an example, one current colleague, Eric Davis, was a civilian college professor at a U.S. Air Force base in Korea. He reported that as much as 30 percent of the floppy disks bought at the Base Exchange contained prepackaged viruses. This was very surprising, as the disks were delivered from the manufacturers in tamper-proof, shrink-wrapped boxes and were assumed to be virus free. He also indicated that some airmen were getting in trouble for electronically importing games from the Internet to their workstations. The virus problem was endemic.[26]

Malicious codes constitute a broader category than viruses. They may include worms, Trojan Horses, stealth codes, bombs, and a number of other nefarious programs that can adversely affect a computer system. The infection may be launched through hardware, software, or firmware, usually hidden in undocumented areas of code that exist in most programs. Entry may be made at almost any vulnerable point in the system, including the computer terminal, the local area network, servers, routers, phone lines, disks, CDs, or even the power supply.[27]

Another IW subject that has long been hidden from view is *perception management*. While everyone knew that perceptions played a major role in shaping the outcome of conflict, the very topic was taboo. It smacked too much of manipulating people in ways that evoked questionable legal issues—issues no one wanted to address. In a 1989 landmark paper, then-Commander Richard O'Neill, a student at the U.S. Navy War College, wrote "Toward a Methodology for Perception Management."[28] It outlined the process necessary for perception management. O'Neill made a compelling case; a briefing based on the paper made its way to the top leadership. They acknowledged that O'Neill was right and told him to bury it.

In preparing for conflict, perception management is essential. The adversary should be led to believe that he is vulnerable and will lose if war is initiated. Potential partners must perceive the cause as just, and victory as assured. The civilian population and political leadership must also perceive the cost as worth the effort. How all that is accomplished is extremely tricky. John Petersen noted that in Desert Storm, "both the military and the press were working hard to manipulate information."[29] Petersen went on to quote General Colin Powell as stating, "You can win the battle or lose the war if you don't handle the story right."

To deceive the enemy is a fundamental tenet of war. However, it was not until 1994 that doctrine on deception was published, and that was in response to the burgeoning field of command and control warfare.[30] The target of the deception is the enemy's decision-making processes. This may be done by directly influencing the leaders, or by manipulating the beliefs of the people who must support them. Colonel Richard Szafranski, in addressing IW theory, rightly stated, "An aim of warfare has always been to affect the enemy's information systems." His considerations for attacking include "every means by which an adversary arrives at knowledge or beliefs" in that context.[31] Szafranski carries that argument to a logical conclusion, *targeting every element in the epistemology of an adversary*. This means attempting to undermine the organization, structure, methods, and validity of knowledge of that adversary. Deprived of valid information, a means to evaluate information properly, or a stable and reliable mechanism for decision-making places the adversary at great risk. Properly constructed and executed, such a sophisticated IW campaign could persuade the opponent's leadership to accept and abide by our will. Imposition of will is, after all, the objective of war.

There always will be constant tension between the military and the media. Great care must be taken when developing a compelling story, to insure that the basic facts are true. Lies and partial truths will be found out. To be effective in the long run, it is absolutely necessary to be well grounded in facts. Spectacular stories may have immediate impact but, when found to be false, they do far more damage to institutional credibility. For instance, in the early days of Desert Shield, stories abounded about babies being carelessly thrown from incubators by ruthless Iraqi soldiers. These tales were even quoted in the hallowed halls of Congress. Later, it was learned they were part of an orchestrated Kuwaiti public relations campaign designed to gain American sympathy quickly.

We know that world leaders obtain much of their information from CNN. In major campaigns, other news organizations follow their lead. Employing perception management techniques includes gaining the willing support of the major news organizations. This is a tedious process that must be developed over many years through trusting relationships. Reliance should be placed on seasoned reporters who have developed a sense of responsibility, ethics, and loyalty. Unfortunately, too many young reporters believe that all institutions are inherently dishonest and the only role for the press is that of an adversary.

In addition to the commercial media, there are many other ways to influence the perceptions of an adversary. Psychological operations units have a wide variety of tools, including radio and television stations, plus leaflets and other printed material. The adversary may be presented with a false picture of the developing battlefield. For instance, during World War II, a phantom army was fabricated near Dover, England. Through a complex scheme of radio transmissions and troop movements, the German High Command was led to believe that the Normandy invasion was a feint, while the real attack would come at the narrowest point on the English Channel, Pas de Calais. The brilliantly executed ruse kept German reserve armored divisions away from the vulnerable Allied forces fighting to establish a beachhead. In Desert Storm, Saddam Hussein was made to believe that the coalition forces would attack straight north into his main defenses. Blinded, his troops were not aware of the powerful tank force that had quietly maneuvered to the west in Operation Left Hook. When the attack did come, the Iraqi guns were still facing in the wrong direction.

Subtle variations of deliberate perception management occur daily. The Internet has become a major source of information for people all over the world. When it is necessary to learn about a new area of interest, analysts frequently turn to the Net and do a topic search. No matter what topic you target, usually a very large number of responses are reported. The analyst must then sort through an electronic labyrinth in search of the desired data. Unfortunately, many of the databases are congested with outdated, useless, and even intentionally false information. Lack of adequate source identification makes assessment difficult. To add to the confusion, sophisticated programmers have learned how to get their message to appear in data searches on unrelated topics. Anyone who has run a search can attest to the weird entries that are listed as high correlation to the subject being investigated. The programmers are conducting perception management, and you are their target.

The fact is, perception management has become a well-honed, fine art in the civilian sector. We call it politics! Advisers carefully check the pulse of constituents and design strategies based on perceptions that are deemed desirable. Truth aside, candidates provide statements that appeal to the audiences, and spin doctors follow up to insure that the message reverberates effectively with different groups. The analogies to politics may be one of the reasons extreme caution is exercised when addressing perception management capabilities from a military perspective.

Perception management is ubiquitous in today's society. Our perceptions, and those of our leaders, drive national policy, programs, and actions. The same is true for our adversaries, and the target of IW is the human mind. The only logical solution is to acknowledge perception management and execute it well. Of all the IW issues, perception management is the most demanding, contentious, and necessary.

As we have seen, information warfare is a complex problem, one that is not well understood. In addition to offensive and defensive aspects, IW does have a *deterrent* component. It is here where American strength may lie. In the amorphous gray areas between peace and conflict, there is an opportunity to apply non-lethal force selectively to prevent escalation of conflict. Information warfare is an ideal tool for sending a very strong message to potential adversaries. That message is that we have the capability, intent, and will to use force—Accede to our demands!

Though complex issues arise with IW, the reader should not assume they are overwhelming. While many will attempt to misuse information systems, it is through dissemination of information that democracies gain strength. Simple protective procedures will go a long way toward preventing IW attacks, even at the criminal level. We must learn to protect our information systems just as we protect our personal valuables. Vigilance and common sense will accomplish most of the task.

11

BIOLOGY

The smallest mistake could be lethal.
—*Tom Clancy,*
Executive Orders

The political pundits had been wrong. The intelligence community had been wrong. Virtually everyone who prognosticated the swift downfall of Saddam Hussein, following the Gulf War, had been wrong. Despite factional conflict, attempted coups, and economic isolation, Saddam sagaciously, ruthlessly, and tenaciously hung on to power. Thousands, including many top military officers, had been shot for real or perceived treason. To limit retribution, their family members often joined in their fate. The very unfortunate ones endured unspeakable torture—burned, broken, and sometimes slowly dissolved in acid—before finally being released from their earthly bonds.

Even members of Saddam's family were not immune. Certainly not Lieutenant General Hussein Kamel Hassan and his brother Lieutenant Colonel Saddam Kamel Hassan, who were both married to Saddam's daughters. General Kamel, himself a brutal killer responsible for tens of thousands of deaths to keep his brother-in-law in power, defected with the colonel and their wives to Jordan in August 1994. Then they foolishly, and voluntarily, returned to Iraq, expecting to be forgiven. Saddam's sisters were granted immediate divorces and the brothers executed—some say by the chief of state himself.

For nearly ten years following Desert Storm, Iraq continued to defy the UN mandates that terminated conflict. Despite being thoroughly beaten on the battlefield, Saddam Hussein continued to maintain sufficient power to be a regional threat. Not only was he allowed to preserve a critical mass of armored forces during the war, but immediately thereafter he began an aggressive program to restore his conventional military might. At every turn, he confronted and confounded the UN inspectors. Hiding evidence of the extent of his nuclear, chemical, and biological programs, Saddam maintained the capability to launch weapons of mass destruction on his neighbors throughout the Middle East. Periodically, he would maneuver his forces in a threatening manner. These shenanigans would be met with countermeasures by U.S. and other UN forces—but at a price. Every time troops had to deploy to the Gulf region in response to inferred threats, it cost money for the movement, and vital training time was lost. More importantly, public support gradually declined as people grew weary of being told repeatedly that something serious might happen. Additionally, international cooperation began to wane. The French withdrew their pilots and took the lead in pressing for an end to the trade embargoes. Other nations tired of involvement in this messy situation that seemingly had no end.

Every so often, in an audacious action, Saddam would push too hard. His troops would be moved into threatening positions, or air-defense systems would acquire UN fighter aircraft patrolling the northern or southern no-fly zones. When serious infractions occurred, cruise missiles were launched against selected targets. This was the case on 27 June 1993, when President Clinton authorized a Tomahawk strike against the headquarters of the Iraqi Intelligence Service near Baghdad. Saddam had been caught attempting the 12 April 1993 assassination of ex-President George Bush while he was on a trip to Kuwait, which definitely exceeded tolerance limits. Although sustaining some physical damage, Saddam would endeavor to turn media attention to miscalculations, such as an errant SLCM missing the target and causing civilian casualties.

As Saddam's activities began to increase, it was determined that additional measures should be taken to reduce the regional threat. Obtaining international concurrence and support for an extensive missile or bombing campaign was out of the question. Still, something had to be done.

It had long been known that a technology existed that was tailor-made for the occasion. Military forces rely heavily on petroleum products. The technology was biocidal-resistant microbes that devour oil. Not only was the technology available, it is the treatment of choice for reducing oil spills on the high seas. As an oil-producing country, Iraq had vast amounts of petroleum available. However, the refined oil that was supplied to the military was stored in segregated, guarded areas. Once inoculated, these oil supplies would spread contamination throughout the fleet of armored vehicles, into other military vehicles, and onto aircraft. As the contamination infected more and more vehicles, it would become increasingly difficult to clean the engines adequately and return them to normal functioning.

By placing the agent directly into the storage facilities, contamination would occur throughout the entire storage, distribution, and vehicle systems. The process would be slow, taking several months before the maximum damage was achieved. Insertion could be surreptitiously made by Special Forces teams. Only a small amount of the agent was necessary to inoculate each storage area. It would be extremely difficult to detect until substantial damage had occurred. The nature of the agent and insertion techniques were such that no identifying fingerprints would be left behind. Unless the A-Teams were actually caught on the ground, there would be nothing for CNN to display.

By restricting attacks to military supplies, there would not be secondary or tertiary casualties. With the mission prepared to launch, the President reviewed the "finding" and picked up his pen.

What he chose to do in this fictional account remains a mystery. The scenario is plausible, the technology real, and the political problems abundant. While the United States and its allies may choose not to use an offensive biological weapon, we must understand that some of our adversaries will not be so constrained. That issue has been mentioned earlier but bears repeating. The biological weapons threat is one of the most serious we will face for the next century. In fact, advances in biology will bring about some of the most dramatic social and technical changes humans have seen. The tip of the iceberg is shown with the issues such as cloning, biocomputing, and biomechanical microrobots. These advances will require us to reevaluate how we think about biological organisms. Their potential is so great that Gregory Benford, a biochemist from the University of California–Irvine, noted that the twentieth century was dominated by physics, but the next hundred years will be known as "the Biological Century."[1]

Before continuing, it is necessary to address two caveats. First, the biological organisms I will address as non-lethal weapons do not attack humans. They are not diseases or pathogens. Current doctrine even restricts the definition of *biological warfare* to "the use of disease to harm or kill an adversary's *military forces, population, food, or livestock*."[2] This limited definition fails to take into account the actual breadth of the BW threat. To comprehend the full extent of the possibilities, it is necessary to understand and include both antipersonnel and antimatériel BW agents in our thinking. The devastating effects of antipersonnel biological warfare deserve to be acknowledged. They are, undoubtedly, the poor nation's weapons of mass destruction.[3] Pound for pound, they are far deadlier than chemical weapons. One estimate states that ten grams of anthrax could kill as many people as a ton of sarin, the nerve agent employed in the 1995 Tokyo subway attack.[4]

Second, we are only going to address naturally occurring organisms, not genetic engineering. While exotic organisms could be developed with today's advanced biotechnology, there is really no need to do so for military purposes. In the antipersonnel field, years of research have shown that anthrax is so effective and persistent that tailored bugs are not worth the effort from a military standpoint. *Bacillus anthracis* is sufficiently pernicious and tenacious that the island of Gruinard off the coast of Scotland, which was used by the British for biological testing in World War II, is still uninhabitable.[5] The British are not alone. Russia has a similar problem from field-testing anthrax. And, were that not enough, the Soviets designed a strain of anthrax that could resist all known vaccines and antibiotics.[6]

While there is no military need for new exotic weapons, research applicable to their development continues. Paradoxically, it is conducted for the purpose of curing cancer. As an example, the deadly biological agent ricin, used by the KGB in London to assassinate Georgi Markov, a Bulgarian defector, is now being engineered to fight leukemia.

Building upon research that showed that blast cells of acute myeloblastic leukemia possess high affinity receptors for the hemopoietic growth factor, granulocyte macrophage colony stimulating factor (GMCSF) has been chemically cross-linked to ricin by protein conjugation. This tethered poison can then enter the leukemia cells by means of the high-affinity GMCSF receptor, where ordinarily it would not be able to gain access, and then kill it. The bifunctional conjugate combines the specificity of GMCSF with the lethality of ricin.

In other systems, antibodies conjugated to ricin can be directed against molecules that are unique to tumor cells. This specificity kills tumor cells without destroying normal cells.[7] In cancer research, such interventions are known as *magic bullets.* Of course, to make this benevolent research a sophisticated biological warfare weapon, only the cells targeted need be altered. Targeting is not a trivial research task, but it can be done.

Biological warfare has a long and nasty history. The Romans first used biological agents in the form of dead animals contaminating water supplies. During the sieges of antiquity, bodies infected with bubonic plague were catapulted over the ramparts to spread illness among the fearful inhabitants. It has been suggested the Black Death, which decimated approximately 25 million people in Europe from 1347 to 1351, may have spread from such an attack on the walled cathedral city of Kaffa located on the Crimean peninsula of the Black Sea. Now known as Feodosiya, Ukraine, Kaffa was a successful Genoese trading port besieged by the Tartar forces of Kipchak khan Janibeg. When plague, in its three forms, bubonic, pneumonic, and septicaemic, broke out among the troops, the Mongols retreated, but not before bestowing their parting gift: a few infected bodies. Encirclement broken, some of the Kaffa merchants immediately sailed for Constantinople, thus spreading the near 100 percent fatal disease to the major trading routes of the known world.

Closer to home, the Native American population living at Fort Carillon—better known under its new name, Fort Ticonderoga—were the victims of a British BW attack during the French-Indian War. In what appeared to be a benevolent gesture, the British provided the Indians with smallpox-contaminated blankets. Once the defenders were weakened by illness, the fort quickly fell.[8]

In the intervening years, crops, animals, and people have all been attacked. During the Cold War, the United States participated in vulnerability testing against unsuspecting citizens in San Francisco and New York. The biggest experiment involved spraying *Serratia marcescens* in San Francisco to see how this supposedly benign agent would spread. The test killed at least one person and infected a number of others. In 1966, a portion of the New York subway system was subjected to *Bacillus subtilis* testing. The results demonstrated that an agent released at one station could be distributed throughout the entire system, moved by the winds created by the trains.[9]

In the 1970s, the then Soviet Union employed Yellow Rain against the Hmong tribesmen in northern Laos. The name was coined by Sterling Seagrave as a title for his book. T-2 Mycotoxin was manufactured in Soviet BW laboratories and dropped from aircraft onto the indigenous jungle people. While the documentation remains classified at code-word level, the BW agent was definitely not bee feces, as proposed by one Harvard professor.[10] More recently, Saddam Hussein used chemical and possibly biological weapons against his own people, killing thousands of rebellious Kurds in Iraq's northern provinces. Ironically, it was the aforementioned Hussein Kamel who was the mastermind of the Iraqi biological and chemical warfare campaign.

In addition to anthrax, people now live in fear of extremely virulent diseases. The mere mention of the dreaded *Ebola virus,* named for the town in Zaire where it was first identified in 1976—evokes fear around the world. Ebola causes hemorrhagic fever and has a fatality rate as high as 88 percent.[11] The mysterious Hanta virus that leaves 60 percent of the infected patients dead only a few hours after the onset of symptoms first appeared in the United States in 1993.[12]

In recent years, books (*The Andromeda Strain, The Hot Zone,* and *Executive Orders*), movies (*Outbreak*), and television programs have done their utmost to terrify us with tales of unstoppable viruses. These invariably emerge from the depths of the African jungles or from the nefarious workings of rogue scientists, and spread like wildfire. While often exaggerated, these stories are grounded in reality and point to the catastrophic consequences of unbridled biological warfare.

For true terror, consider what might happen if genetic engineering were conducted on one of the five known strains of Ebola. The reason Ebola does not spread extensively is that the virus is transmitted through bodily fluids. Once the disease is recognized, containment is manageable. However, if the current strains were intentionally conjugated to an airborne host or an influenza receptor, transmission rates would jump dramatically.

Even without human meddling, natural mutations are taking some ominous turns. In 1982, *E. coli* 0157:H7 was first identified in the laboratory and remained almost unheard of for a decade. By 1993, it was called "one of the most dangerous food-borne bacteria known to mankind" and had sickened millions, while killing thousands, of people. For this disease, there is no vaccine and no treatment.[13] Additionally, overuse of available medicines has fostered the evolution of new antibiotic-resistant—and in some cases totally immune—viruses. Similarly, untreatable tuberculosis is now emerging in hot zones around the world. The estimates on the number of such cases range from 2 percent to 14 percent of all the cases in the world.[14] Bubonic plague resistant to drugs has also recently emerged. Our cavalier attitude, demonstrated in devices such as no-sweat pills, has produced horrendous viruses that are nearly unstoppable.[15,16]

Besides humans, plants and animals are lucrative targets for BW attacks. U.S. agriculture represents an extremely large and vital industry, and it is vulnerable to a wide variety of plant and animal viruses. Lieutenant Colonel Robert Kadlec, writing at Air University, noted the disaster that could befall the United States if agriculture was attacked.[17] Ironically, it was the United States that conducted extensive research in methods to destroy crops. In fact, the United States has recently been accused of instigating such an attack against Cuba. The claim is based on the authorized flight of an American crop duster aircraft over Cuba en route to Bogota, Colombia, on 21 October

1996, and a subsequent infestation of thrips palmi, a native Asian bug that is extremely aggressive in destroying crops. A Cuban pilot reported seeing a white mist coming from the plane. Although the accusation was immediately denied by the United States, it demonstrates the difficulty in establishing causal relationships of biological agents.[18]

Unfortunately, a BW attack in any of these areas is a low-cost option. By one estimate, a single ounce of some biological agents, costing about one dollar, could contaminate an area the size of a square kilometer. The Office of Technology Assessment stated that a large BW arsenal could be developed for as little as $10 million. At that price, any adversary, even wealthy individuals, could acquire strategic capabilities.[19] A basic understanding of the historical perspective and future threats from biological warfare against people or other living organisms can be gleaned by reading the military papers of Lt. Cols. Terry Mayer and Robert Kadlec.[20,21]

Beyond the horrific confines of antipersonnel BW lies a technical and ethical conundrum. When I first began exploring *antimatériel* biological technology, I was amazed at the range of vulnerable targets. Based on advances in bioremediation, the materials that could be degraded included petroleum products, explosives, plastics, adhesives, and even metals. The list of possibilities kept expanding. Finally, we came to understand that there was almost nothing in the world that some organism will not consume.

The growing amount of trash is a global problem, and remediation is essential. Especially in technically advanced societies, we tend to throw away vast quantities of materials. In fact, for many products we even have designed obsolescence with little regard for methods of disposing of the unwanted items once their period of usefulness has passed. Developing nations are contributing to the problem as income increases and they attempt to emulate the developed nations that generate greatly disproportionate amounts of refuse. Contamination of the oceans inhibits dumping at sea, and more and more lesser-developed countries are taking the NIMBY (not in my backyard) approach to accepting waste materials. Therefore, how to deal with the ever-mounting volume of nonrecycled trash is indeed an issue of monumental proportion.

Biochemists all over the world are working on the development of organisms that can reduce the amount of refuse necessary to store. Again, the biological paradox is that the same technology used for peaceful purposes comprises a very significant threat to our national security. Only the intent of the user changes. Remember, the designation of what constitutes *trash* is arbitrarily assigned after some unspecified period of usefulness. The physical composition of the item doesn't change upon retirement; only the way we think about it is different.

The very bad news is that biological agents are relatively easy to acquire. One way is simply to buy them. Remember, Saddam Hussein ordered his supply of anthrax from a mail-order house in the United States. Another approach is to cultivate the desired organisms. To do that, one need only find a place where the target substance (oil, plastic, metals, etc.) has been deposited in the ground for a period of time. The task then is to locate the organisms that have grown in that area and have adapted to the material as a nutrient. The rest of the process is a matter of breeding for the most beneficial characteristics. Examples of such characteristics might include tolerance to weather, resistance to biocides, or accelerated reproduction. Once organisms are isolated that display the desired attributes, they then are concentrated through selective breeding until they are suitable for use in attacking the designated target.

Pat Unkefer, the scientist leading the bioremediation effort at Los Alamos, estimated that a competent high school science student could learn how to collect and cultivate biological organisms with about two hours' training.[22] It can be done with very modest equipment, and the procedures can easily be conducted in a garage or basement. These activities would be virtually undetectable. The techniques are also readily available in books on biology or from the Internet.

There are natural boundaries on the growth of organisms. Organisms tend to grow exponentially, then stop once the nutrients have been consumed. Just as there is no universal solvent, there is no universal organism that can eat everything. As Unkefer puts it, "There is no bug that can eat Cleveland." While some might be willing to sacrifice Cleveland, her point is that biological organisms mutate slowly, and they do not adapt to totally unique nutrients in a short period. Were that the case, everything in the world would have been destroyed by now.

Still, fears persist that if an organism is introduced to degrade one crop, after accomplishing that mission it will take on another. This was the case when the alteration of coca was discussed. However, advances in genetic engineering provide relatively safe options. For instance, it would be possible to reduce dramatically levels of cocaine, the crystalline alkaloid $C_{17}H_{21}NO_4$, which produces the "high" that leads to addiction. The approach would be to genetically engineer the coca plant with a key enzyme inhibitor that blocks natural biosynthesis of cocaine. Then, by cotransection with an herbicide, a pure population containing only the new strain would survive after the fields were sprayed. This technique provided a viable alternative to crop eradication, which has proven to be nearly impossible. However, opponents suggested that after altering coca, mutation affecting tomatoes might be initiated—a claim without scientific foundation. As with lasers, biotechnology is an emotional issue. This particular effort was abandoned for political, not technical, reasons. Still, the quest for new biological agents will be pursued—for legitimate purposes.

The breadth of materials that are susceptible to biological attack is awesome. To those outside the field of biochemistry, bugs that eat metal seem like a stretch of the imagination. However, the 1998 phenomenal blockbuster film *Titanic* introduced many people to this concept as documentaries noted that iron in the hull of the ship was actually being consumed by microscopic organisms that inhabit the ocean depths.

There are more intentional applications of such biochemistry. For instance, biological organisms are actually used for mining certain rare heavy metals, such as platinum and uranium. At New Mexico Tech, an organism was created that consumed gold. These organisms are introduced near the ore and begin to absorb the metal. Being living creatures, they are mobile and move to a specific location where they are gathered, harvested, and the metal extracted. A similar process is used in the remediation of actinides from nuclear waste. Jimsonweed may be grown over plutonium residue and the plant will extract the man-made substance and incorporate it into its body.

Of course, a disposal mechanism for the plant is necessary, as the process does not eliminate the problem; it only transforms it into a more manageable mode. In fact, the process of directed mutation, while having many benefits, also has potential drawbacks. In essence, these processes introduce new organisms into our biosphere. As they are new, we do not fully understand their consequences, so caution must be applied at all times when developing these life-forms.

The example of contaminating a petroleum reserve was one in which a large body of material was subsequently degraded. There are other, similar military targets. Explosives, also called *energetic materials,* can be affected by destroying critical nitrogroups, thus reducing the blast effectiveness. Consider that the degradation half-life, using *P. fluorescens* III on TNT, is about one week. In other words, two weeks after inoculation, the explosives could explode with only 25 percent of the original power. One can imagine the impact of introducing these bacteria at the point of manufacturing. A very small number could have very large effects.

An antimatériel approach that offers great potential is to degrade chemical bonds rather than massive bodies of substance. Plastics offer examples of substances that are often hard to degrade and have led to burgeoning landfills. However, rather than attempting to destroy a plastic object, it is easier to attack the bonding chemicals, known as plastizers. It is plastizers, including *adipates* or *sebacates,* that hold the plastic polymers together. These commercial ester bonds are known to be susceptible to fungi such as *aspergillis.* The approach may not help quickly reduce the volume of landfills, but from a military perspective, it offers a method for attacking enemy matériel. Other vulnerabilities include adhesives, rubbers (natural or synthetic), resins, coverings, joints, metal pins, or insulation. The most viable places to attack and degrade are the critical subcomponents. A small amount of physical damage at key points can prevent an entire weapons system from functioning properly.

Similarly, paints and special coatings applied to aircraft may be vulnerable to attack. Once degraded, reduced radar signals could be enhanced, thus leading to ease of detection. As part of the biodegradation process, minute granules called inclusion bodies made of salt crystals, metals, or plasticlike polyhydroxyalkanates are naturally produced. Such particles could quickly clog filters and introduce abrasives or gums into critical weapons system components.

Larger-scale systems, including critical elements of transportation networks, are also potential targets, as they are vulnerable to rapid, component-specific attack by microbes. As pointed out by Captain James Campbell of the Naval Research Laboratory, such attacks could result in "acceleration of corrosion, degradation, or decomposition of roads and aircraft runways."[23] In this most important presentation, he addressed the range of matériel targets and noted that the next step would be to modify genetically microorganisms that are "expressly focused on degradative capabilities." To ensure that predetermined limits were in place, timed "suicide genes" or other alterations would be employed to establish boundaries of time or space. Captain Campbell suggests that the next step may be to "develop biomimetic chemical systems that reproduce specific degradative capabilities, but without the requirement for living organisms." Since there are no living organisms involved in such methods, such agents would not be subject to existing treaties.

Freeze-drying technology adds another dimension to the use of biological organisms to attack specified targets. In late 1991, microbiology professor Raul Cano at California Polytechnic State University was able to revive living microorganisms trapped in amber for millions of years. Cano's prior work, isolating viable ancient DNA, was the process that sparked Michael Crichton's best-seller *Jurassic Park.* By the time Cano announced his new, validated discovery three years later, his team had revived 1,200 types of bacteria dating back as far as 135 million years.[24] This proved that the dehydrated organisms could lie dormant for extremely long periods of time and then be brought back to life. This has significant implications for weapons applications.

A similar approach would enhance the capabilities of antimatériel biological weapons. Rather than relying on insertion of living organisms, freeze-dried ones could be introduced into the target with a means to revitalize them at a specified time. This would provide time for the agent depositing the weapon to be long gone from the scene. Even years later, the weapon could be reactivated through a time-delay mechanism or remote control. A vial would be ruptured, introducing the bacteria to waiting nutrients, thus allowing the attack on the target to begin. National fingerprints would be almost impossible to identify. It also means that the shelf life of such BW weapons would be extended indefinitely.

Regardless of one's philosophical position on the use of antimatériel organisms, it should be clear that their potential is very great. Biological weapons of many types exist, and they will not disappear, either by wishful thinking or by treaties. They constitute one of the greatest threats to the security of all nations.

Countering the myriad of threatening organisms can be mind-boggling. In fact, some science review boards have ducked the issue by putting BW in the "too tough to handle" category. The situation should not be viewed as hopeless since the lessons of their use are finally hitting home and being addressed. Medical journals and news media are beginning to report openly about the topic. In August 1997, the *Journal of the American Medical Association* warned about the dangers of biological terrorism.[25] Major newspapers have also carried the stories.[26] And as of 25 August 1997, the Army began to prepare civilian emergency teams in large U.S. cities to counter a CW or BW attack.[27] While the threat is very difficult to defend against, measures can be taken to identify and deter organizations that proliferate biological weapons. At a minimum, we urgently need concerted research into countermeasures for the full range of BW agents. Actually, vaccines can be used for many agents, and the consequences of most biological weapons are treatable if quickly detected and identified. Thorny questions about who gets vaccinated and the availability of medicines will generate debate for years to come. For now, the most critical issue is to develop sensors that alert us to inoculation by an adversary and developing civil and military response techniques.

Part III

OPERATIONAL SCENARIOS

"Give me examples of how you would use non-lethal weapons." This is the most common request at a non-lethal weapons briefing. Even after being told about the capabilities of these new weapons systems, almost everyone asks for real-world situations in which they could be used. Every study about non-lethal weapons I have been involved in has started by developing a set of scenarios, then devising weapons that have the capabilities to solve that particular problem. While somewhat limiting the scope of each study, these scenarios provide weapons developers and military planners with concrete circumstances from which they can hypothesize requirements.

Therefore, Part III provides the reader with four vastly different scenarios for the use of non-lethal weapons. They include peace support operations, technological sanctions and strategic paralysis, as well as domestic police encounters. While they are my own works of fiction, they are all based on real-world situations and are all technically feasible, though some of the weapons would require additional development. None of the applications should be viewed as U.S. military doctrine or as policy for law enforcement agencies. Also, they are not the only possible situations in which non-lethal weapons would provide a valuable adjunct to existing systems. Rather, they are designed to allow us to integrate the weapons with possible operational concepts.

12

PEACE SUPPORT OPERATIONS

Peacekeeping is not a job for soldiers,
but only soldiers can do it.
—*Dag Hammarskjöld*

For years, several countries in the Balkans hovered in a twilight zone between conflict and peace. Until the fall of the Wall, personal and group animosities layed dormant, repressed by the overbearing presence of the Soviets. Once the coercive dominance was released, grievances, some which had been festering for decades, were unleashed on the landscape, causing incalculable misery locally and great concern for all of Europe. The Western world tried to make sense of the seemingly inexplicable violence but failed to comprehend the complex factors that precipitated such deeply ingrained, long-standing hatreds. How could people who had coincidentally endured hardships and shared close personal relationships, occasionally intermarrying, instantly dissolve into camps as bitter enemies?

The significance of these events greatly exceeded the boundaries of an isolated geographic area, as it hinted at the very foundations of human nature—truly a frightening perspective. Could it be that man had not evolved nearly as far as we like to believe? Were violence and tribalism innate? Was it only through the application of overwhelming force that humans would behave in a civilized manner? These were questions many did not want to face. Nor were these purely philosophical questions. There was a practical perspective as well. Our collective view of the answers would determine how we approach the issues of national security. If we accepted man as merely a barbaric animal, then the structure of force necessary to maintain freedom would be much larger than if he was judged to be rational and likely to act in his own best interest. Then methods other than brute force could be employed to gain our objectives. However, history is replete with examples of weakness producing vulnerabilities that will be exploited sooner or later.

In the Balkans, periods of quiescence were frequently punctuated with lapses of violence, the sources of which were difficult to trace. There were claims and counterclaims about responsibility for these incidents. Although the "ethnic cleansing" mass murders of the early 1990s had ended, slowly and steadily the bodies continued to pile up. During the month of June, intelligence agencies of several nations reported activities that could lead to renewed open hostilities. In early July, combatants known to be members of provisional units were congregating near the areas in which heavy weapons had been sequestered. In the event that they moved to obtain these weapons, the limited UN peacekeeping force would be inadequate to prevent capture of the weapons or themselves.

On 16 July, communications intercepts indicated that an attack was imminent, probably within the next twenty-four hours. While the recently expanded NATO countries had more resources upon which to draw, most of the newly incorporated nations did not have sufficient experience to function within the complex command-and-control system that had evolved from five decades of cooperation and preparation for war in Central Europe. That limited the response options to nations that had practiced together for many years and participated in peace support operations. Of course, there would be the political hurdles endemic in any multinational operation, most notably, agreeing on objectives and obtaining commitment of forces. Of great concern was the safety of the forward-deployed troops. While they had been invited into the area by all contested parties, technically they were observers and not equipped to enforce peace. Their capture would exacerbate the situation. Should they be threatened with harm or summary executions begin, years of constructive intervention could be lost instantly. The countries providing the peacekeepers would demand use of all force available. Despite allegiance to NATO, they might decide to strike unilaterally in efforts to rescue their people.

Muslims living in isolated enclaves were rightfully worried about their fate. In earlier engagements, several UN-protected territories were abandoned in the face of large, hostile forces. Later, as mass graves yielded their grisly contents, the price paid for untenable threats was made known to the world. As their concern for safety rose, they began to follow a familiar pattern. They crowded together and huddled as close as possible to the UN protectors. This instinctive maneuvering by frightened people placed the peacekeepers in an extremely difficult position. If they stayed, they were subject to capture and becoming hostages. To leave, they would have to extricate themselves from people who were trying desperately to prevent that from happening. The use of lethal force against terrified women and children was unthinkable. To remain unprotected was equally undesirable.

Under the UN mandate, the peacekeepers were authorized to take measures to limit future fighting. However, they could only fight if directly threatened. As their current position was militarily untenable with existing forces, the decision was reluctantly made to withdraw the ground forces. However, before their departure, the NATO forces would render unusable as many of the stockpiled heavy weapons as possible. So as not to alert the provisional forces surrounding them, this degradation would take place silently.

At 0115 hours on the 17th, specially trained engineers entered the compound in which the artillery and armored vehicles were stored. Exercising great care, they placed small, self-adhering vials containing the binary chemicals of superacids at predetermined critical areas of each weapons system. Once emplaced, the engineer twisted a small handle, rupturing the membrane that separated the chemicals. While each of the binary components was strong, when combined their acidity was amplified by orders of magnitude. Within thirty minutes, the superacids would eat through the casings and begin to etch away at any materials with which physical contact was made. The objective was not destruction of the large structures, but rather degradation of key components. In a matter of two hours, small pins would dissolve, gaskets would leak, wires would erode, fuel tanks would develop holes, optics would craze, and many other nodes would cease to function effectively.

In less than half an hour, the storage depot was entirely rigged. Even before the NATO forces initiated the second phase of this mission, the large guns would no longer be able to inflict their massive damage. The aggressive chemicals would be transformed in the process. Thus, the potentially hazardous effects of the superacids contacting humans would be negligible. To reduce the risk of incidental contact further, at 0330 hours water was sprayed onto the weapons, completely dissolving and washing away any remaining superacids. As very small amounts of water-soluble superacids were used—and after toxicity testing—there would be no environmental danger to this controlled application.

In the meantime, the NATO troops prepared to evacuate the area. The route would be in a column following the main roads to the south. It would not be possible to slip out unnoticed. The Muslim civilians were hyperalert for such a situation. The commercial radio and television stations had been carrying stories about the plight of the peacekeepers and increasing tensions in the area. Air evacuation was ruled out, partially to keep the NATO equipment intact, partially for fear of reaching a point at which a small remaining force might be overwhelmed by one side or the other. Also, while NATO maintained air supremacy, there was the possibility that some helicopters might be shot down by small arms fire or by Soviet-made, shoulder-launched antiaircraft missiles known to be in the possession of both sides.

The first major hurdle would be to disengage from local civilians without resorting to lethal force. A combination of tactics was employed. By 0500 hours, the peacekeepers were prepared to move. To prevent massing of the civilians, the forces remained dispersed. Just as the troops assembled at their designated areas, a pair of F-15s came screaming in, scraping the deck. Then, hitting their afterburners, the Tomcats soared straight up into the early morning skies. The thunderous noise shook the buildings and left a clear, unnerving psychological impression—they meant business. In actuality, these two fighters were only a part of the air cap that would cover the troop withdrawal. The message they imparted was that overwhelming firepower was present, and any attack on the departing troops would be decisively and immediately crushed.

Simultaneously, a call was placed to the command post of the provisional forces. They were told that the NATO peacekeepers would be leaving and interference not tolerated. The commanders were also told they would be held accountable for any actions taken against the Muslim civilians left in the enclave—probably an idle, but also an obligatory, gesture.

Separating from the now terrified civilians would be the most difficult task. As a means of distraction, two C-130 aircraft circled a short distance from town and began making an air drop. Under the brightly colored parachutes were bundles containing nonperishable food. It would be difficult for the civilians left behind to find food in the days to come. It was hoped that a substantial number would be attracted to the supplies, thus reducing the people with whom the troops might need to confront.

As the troops began to move, civilians gathered about them and begged them not to leave. The situation was heartbreaking. Women and children were crying and clutching onto anything they could grab. Faced with no good choices, the commander sent word to use a mild form of pepper spray in hopes of driving the crowds back. Some moved, others persisted. As the column worked its way south, a helicopter, specifically configured with an acoustic device, flew to the north. As it cleared the convoys, it began emitting infrasound pulses, directing them toward the crowd below. Those in the beaten zone rapidly scrambled to get away.

On command, the entire peacekeeping force donned their protective chemical gear. Then suddenly, a crop-duster aircraft was overhead, spraying stronger riot control agents. A second crop duster joined in and began clearing the path to the south so that no additional civilians would approach the departing forces. For the few who continued to venture close to the troops, flash-bang grenades dissuaded them.

The mission to retreat was accomplished with no immediate loss of life. NATO had been intimidated into leaving a declared protected zone. Still, they disengaged, leaving no soldier behind to become a hostage. Now the diplomats could try yet again to reinstall peace. It was not a banner day for NATO.

While this account is fictitious, it is a realistic portrayal of the predicaments peacekeepers may face. Though they are undertaken with the best of intentions, sometimes peace support operations lead to unintended consequences. Many of the technologies employed were discussed in the NATO-AGARD study to determine methods by which collateral casualties could be limited during PSOs.[1] That study was based on NATO experience in Bosnia. To assist the study group members in conceptualizing the problems associated with limiting collateral damage in these difficult situations, a number of *real-world* targets were picked for neutralization.

Three categories of targets were selected: personnel, equipment, and infrastructure. For each, a situation that had been encountered during peacekeeping operations in Bosnia was developed. Examples of the target situations include the following: neutralize a mortar crew for at least twelve hours; halt an aggressive crowd; prevent missiles from being fired; prevent a tank from firing the main gun; deny use of a bridge for twenty-four hours; and many others. Of course, the situations were modified to be more difficult than attacking an isolated target. For instance, suppose the missile launcher just happened to be placed next to a hospital, while the bridge was built in antiquity by the Romans and of great historic value and the tank was intentionally hidden in a farmer's barn.

For each scenario, solutions were developed, some with non-lethal weapons, others with low-collateral-damage techniques such as using precision-guided weapons with inert warheads. While they produced a hard kill of the target, without the high-explosive warhead little fragmentation occurred that might dismember unfortunate bystanders. The study concluded that there were no *silver bullets,* no perfect weapons that could be employed in every situation. However, it did show that while it was difficult to attack targets contiguous to sensitive civilian sites such as schools, houses, hospitals, or religious buildings, using precision-guidance systems, several special warheads could be employed to limit the use of enemy equipment or facilities. For instance, adhesive foams would be effective in inhibiting use of the Roman bridge without causing any structural damage, and foams could be sprayed on sophisticated equipment requiring extensive maintenance prior to renewed use. Olfactory insults could be used to keep people out of specified areas, and they would be safe in case direct contact was made with the substances.

The concepts and doctrine for peace support operations are still emerging and even the vocabulary is not yet established. However, it is clear that a wide range of missions are included under the rubric of PSOs. Peacekeeping is the maintenance of a diplomatically established termination of conflict but one in which renewed fighting could take place. Normally, peacekeepers are invited into the area by all parties.[2] Peace enforcement includes the use of external forces to quell outbreaks of violence by parties who have agreed to cease fighting. Mutual consent may not be a part of peace making, and there is an increased likelihood of armed interventions. Preventive diplomacy includes actions taken in advance when a crisis is predictable.[3] Peace-building missions are undertaken to establish an infrastructure in which democratic governments can gain and maintain control over the local situation. Additionally, humanitarian assistance and support to noncombatants in distress may be part of any of the other PSOs, or an independent operation.

Another related peace support mission that is gaining prominence is the evacuation of civilians caught in areas of unrest. Between March and June 1997, U.S. Marines engaged in evacuation operations in Albania, Zaire, and Sierra Leone. The year before, they were called to Liberia and the Central African Republic. While primarily present to rescue U.S. citizens, the Marines have also helped hundreds of foreign citizens escape the ravages of war.[4] Since these situations are very volatile and dangerous, the troops and transports are fully armed with lethal weapons. If fired on by snipers or militia groups, they have returned with deadly fire, from both ground troops and supporting helicopter gunships.

Conducting the extraction requires the insertion of a force powerful enough to adequately gain control of the local situation. While some evacuations may be supported by troops already in place, such as at the U.S. embassy, others will not have anyone prepositioned. Forced entries under highly ambiguous circumstances are very tricky. Inadequate force met by a violent response could result in the loss of lives of troops as well as the lives of those they are attempting to rescue.

Conversely, use of excessive force can claim many innocent lives and adversely impact relations for a long time. That can be disadvantageous on an economic level. Consider the situation in Zaire, now known as the Congo.

The area is known for its rich mineral deposits, but the instability in the capital put foreigners temporarily at risk. The long-term consequences of application of force were very important if mining was to continue.

In addition to lethal force, these units need non-lethal capabilities to be able to establish and maintain control of prescribed areas. Since they may need to inhabit those areas, at least temporarily, acoustic weapons, lights, and lasers may be appropriate. To establish corridors for travel, smoke and other obscurants can block observation of the operation by potential adversaries. Infrared vision devices allow the soldiers to see quite clearly while the indigenous forces, lacking sophisticated equipment, cannot.

However, as in the Balkan situation described previously, there are often attempts by traumatized local civilians to accompany the rescuers to safety. One of the most difficult tasks in evacuations under pressure is to determine whom to take. Since the situation is analogous to civil war, it is the responsibility of that nation to care for its own citizens, and often a nation is not capable of doing so. In the chaotic conditions of major unrest, people may not have time or opportunity to retrieve their passports or other documentation of citizenship. As there are no obvious physical characteristics that identify Americans, sorting may take place after departure. However, there is a direct correlation between the number of people to be evacuated and the resources necessary to bring them out. For troop safety, the smaller the number, the better. Also, they must carefully regulate the number of people boarding each helicopter. There have been too many tragic incidents in which overloaded rescue helicopters have crashed. Those who remember the fall of Saigon will never forget the graphic films of people, in acts of abject desperation, attempting to hold on to any appendage beneath departing aircraft. We watched in horror as they plummeted to their deaths as the strain to maintain a grip exceeded the limits of human endurance.

During evacuations, very hard decisions must be made in real time by young troops. Even though fleeing people may not be armed and have no intention of harming the troops, they can precipitate very dangerous situations, such as attempting to rush the aircraft. Preventing unauthorized civilians from boarding departing helicopters by use of deadly force is highly undesirable—and makes for very bad press. Many of them may be women and children whose only goal is to reach a safe area. Such situations beg for non-lethal alternatives to be made available to help protect troops and civilians alike.

Humanitarian missions bring another unique set of problems, especially if they are in the category of disaster relief. In such a situation, there is no enemy. There are, however, frightened and confused people motivated by their most basic instinct: survival. Providing food to starving people can be very hazardous. This is exacerbated if people perceive that supplies are inadequate to feed everyone or that the distribution processes are unfair. Perception management and psychological operations are essential for the safety of the supporting personnel from both military and civilian aid organizations.

Additionally, security of base camps and distribution sites may be difficult. Undoubtedly, the military troops involved in the operation will be better fed and better equipped than the people they are helping. Even field rations are likely to be coveted, as they are oftentimes of a much higher caloric content and quality than the relief rations being distributed. Soldiers may experience their own psychological trauma when they realize that every morsel they eat is badly needed by the population they are helping. Nonetheless, the health and strength of the troops must be preserved if they are to accomplish the mission. Likewise, their physical safety must be maintained. Non-lethal weapons are an essential part of the inventory for humanitarian missions.

Even peace support operations have a dark side. Shortly before midnight, Major General Bohgdan Bagdonovich left the restaurant where, by Bosnian standards, he had dined rather elegantly. Considering European custom, this was an early retirement. But the general had meetings the next morning and he wanted to be mentally alert. Followed closely by a car loaded with bodyguards, Bagdonovich would be driven the short distance to the next village, where he resided temporarily.

Three months earlier, an international tribunal of judges in the World Court in The Hague[5] *had issued an arrest warrant for General Bagdonovich for "crimes against humanity." The testimony of survivors and lower-ranking soldiers had been riveting. In addition to authorizing the extermination of civilians trapped in captured villages, Bagdonovich had personally executed several captured soldiers—"to demonstrate appropriate resolve," he had told the soldiers. He had also participated in other "fruits" of victory, the raping of several young girls. After the girls were turned over to other soldiers, who relentlessly took turns, their badly beaten bodies were unceremoniously dumped in a dung heap.*

Since his indictment, Bagdonovich continued to be a public presence, almost taunting the rest of the world. In his home country, he moved with impunity and even allowed the press to photograph him. On occasions he even had the audacity to travel through UN-controlled checkpoints, where they failed to recognize him. Being early summer, the sun had set around 10 P.M. and would rise again in a few hours. Tonight, the general's whereabouts had been determined through signals intercepts by the U.S. National Security Agency and assistance from Polish intelligence agents, who narrowed the search and confirmed his arrival at the restaurant.

Halfway along the eight-kilometer journey waited Captain Harmon and five other highly skilled Special Forces soldiers. Their assigned mission: Capture General Bagdonovich and bring him into UN custody—alive! The Special Forces team had successfully been inserted two nights earlier and remained hidden, observing the ambush site. Tonight they received the execution order and moved into position. Tipped by an agent, the team was fully prepared when Bagdonovich's cars proceeded up the road with their headlights on high beam. One hundred meters before the ambush site, two small charges were detonated, destroying the front tires of the trailing car. The lead car proceeded on, not aware that the attack had begun. Seconds later, several loud explosions heralded the formation of an abatis created by interlocking fallen trees.

As Bagdonovich's car attempted to flee while driving on the rims, the engine block literally was impaled by a rocket-driven, depleted uranium rod, thus ending the evasion. Immediately, a whistling sound was heard. As the guards strained to see what was making the noise, brilliant pyrotechnic munitions illuminated the area as if someone had suddenly turned on the midday sun. Flash-blinded, the guards would be useless in protecting their ward. As they stumbled from the car, they were ensnared by nets and quickly subdued. The lead car was also engaged with optical munitions, followed by rapid-firing flash-bang devices. Believing they were under direct attack, the guards fired wildly but were no danger to the snatch team.

His eyes revealing nothing but overloaded photoluminescence, Bagdonovich could not see the approaching soldier. As he was grabbed and thrown down roughly on the back seat, he felt a stinging sensation as the needle penetrated his leg. Then the world dissolved into a sea of black. A hood was placed over his head and Bagdonovich was bagged. Two members of the SF team dragged him from the car and placed him on a waiting litter. They would have to carry him several hundred yards to the exfiltration point. The helicopters were already inbound and would be on station in less than ten minutes. Quickly, the team covered the distance to the open field and signaled the waiting chopper to land. In less than fifteen minutes from first contact to liftoff, the successful snatch mission resulted in one of the most wanted war criminals being brought into UN custody.

The notion that special operations soldiers would function like Texas Rangers of the Wild West era may rankle most military officers. But, for all of the debate about the appropriateness of U.S. troop involvement in peace support operations, nothing is more acrimonious than the issue of apprehending war criminals. Such missions would call on them to penetrate the hostile territory, where the accused is at home, surrounded by supporters and guards. They would then have to snatch the prisoner—alive—and all make it back safely to deliver him or her into the hands of justice. Of course, when the first such raid was conducted, they were partially successful. The NATO forces involved were able to capture Milan Kovacevic, the primary target. However, others around him were not so lucky when they resisted arrest.[6] Later raids in the area by a NATO contingent of British and Czech resulted in the confiscation of more than 2,500 weapons, plus ammunition, grenades, and explosives. The mission was again creeping.[7] As such actions occur, it is hard to remember a UN motto that goes, "A UN soldier has no enemies, only parties."[8] Actions such as these have resulted in what some term "wider peacekeeping." It acknowledges the reality that at times forces will be called upon to bridge conventional peacekeeping and limited warfare.[9]

Many decry the open flouting of international law by factional leaders whose despicable crimes against humanity make serial killers like Jeffrey Dahmer and John Wayne Gacy look like choirboys. For their participation in systematic torture, mass murders, and "ethnic cleansing," voices around the world cry out and demand they be brought before a court of law, there to be held accountable for their heinous acts. Everyone agrees with the need for accountability, but just how to accomplish the task is at issue. Opposite sides of this contentious discussion were taken by Secretary of State Madeleine Albright and Secretary of Defense William Cohen.[10] Albright wanted to authorize the abduction operations, while Cohen thought it inappropriate for American forces. For soldiers assigned such a mission, the physical and political risks are very high. Use of excessive force, resulting in the death of the accused or the wrong collateral people, would be construed as worse than failing apprehension. Ironically, if any of the special operators die in the process, that is seen as "just part of the job."

Political aspects aside, such missions are difficult but technically feasible. In order to make the arrest and return with a live prisoner, non-lethal weapons are a valuable—if not essential—adjunct. It must be remembered that enforcers are executing an arrest warrant, not a death warrant. On a prisoner snatch, known to the troops as "bagging," the use of incapacitating chemicals can be considered. Since the body weight and health condition of the targeted individual are known, exact drug doses can be prepared that have a high probability of rapid enervation at minimal risk of serious injury. While the sought-after individual is known, who else might be around him or her at any given time cannot be predicted. Here again, authorization of force must be carefully considered before embarking on the mission.

This chapter illustrates only one of the more contentious aspects of peace support operations. There are many more mundane actions that could have been described. They might range from protecting humanitarian relief efforts, maintaining refugees, and reinstalling civil order, to active peace enforcement. The biggest threat to peace support operations is the number we are attempting to undertake. Many believe that our forces are already spread too thin and at the expense of the primary mission of the military—national defense. The Army, in *Vision 2010,* notes that between 1990 and 1996 they deployed forces on 25 missions, a tremendous increase over previous years. The question is whether or not we can continue at that rate without degrading our combat capability.[11]

Obviously, we will continue peace support operations. The amount of force necessary will be dependent on each situation. Worse, those situations are likely to change—possibly abruptly. Therefore, troops involved must have the right tools for the job. In many peace support operations, that includes both lethal and non-lethal weapons.

13

TECHNOLOGICAL SANCTIONS

If you can keep your opponent's nation intact,
then your own nation will also be intact.
—*Jia Lin (Tang Dynasty)*

November 21 was TS Day. That stood for technological sanctions day, the day the United States, backed by a UN resolution and general support of the world community, initiated actions against Bechistan. By the end of the day, it was clear to the Bechistan government leaders that the UN meant business and was prepared to use force to achieve its objectives. First, the telephone system had been blocked, then local telecommunications systems failed to transmit over their normal areas and only external broadcasts were seen and heard. Shortly thereafter, computer systems linked to outside networks crashed, one after another. Degradation of other major infrastructure systems would follow!

For several years, the mythical southwest Asian country of Bechistan posed serious problems for both her neighbors and the rest of the world. As a transit point for oil, it was always questionable as to whether they would continue to let the product flow to the seaports or, once there, to allow it to be loaded on waiting tankers. Two months previously, they had threatened to send forces into the disputed border areas claimed by both Bechistan and Pakistan. This aggressive stance was precipitated by discoveries of rare minerals in the remote desert area. Ground-penetrating sensors, probing down far below the surface from low-earth-orbit satellites, had provided the first indications of the deposits. On-site research verified the high probability of resources that would provide the owner with wealth for years to come. It was a tempting target of opportunity. Although Pakistan had larger forces, it was preoccupied with events near Kashmir, a perennial flash point between it and India. Tensions on that northeastern border area had grown to the point where Pakistan deemed it necessary to keep the vast bulk of its huge armored force in the eastern states. India, being an acknowledged nuclear state, demanded Pakistan's full military attention. Long suspected of possessing nuclear weapons, the Pakistani arsenal was far short of that of either China to the north or India to the east.

Bechistan, while not normally aligned with India, had signed an agreement with that country to provide trade and a degree of security. Political scientists argue about the alternating preeminence of culture versus economics. In this instance, basic greed and economics was chosen over cultural identity. However, this alliance was complicated by a residual tie between Bechistan and China, a traditional adversary of India, and it was one with more nuclear weapons than any other nation in the vicinity. In short, the situation was a classical problem for the United States and other Western countries with interests in the area. While regional stability was necessary for the good of the global economy, the balance was very delicate. The United States maintained a perception of ability to use force in the area, but the reality was that it would be very difficult to accomplish. The geographic distance represented a major problem. The physical presence of the United States was limited to a large supply base on the island of Diego Garcia, and whatever level of intimidation the navel fleet could muster.

Therefore, government leaders in this area of the world paid only modest attention to threats of military intervention by the United States, or even other European nations. Iraq's Saddam Hussein clearly demonstrated this when he not only invaded Kuwait, but then stood toe-to-toe against the UN coalition forces. The results of Desert Storm are well known. Militarily, it was a disastrous defeat for the Iraqi forces. While it was not 100,000 killed, as many believed, tens of thousands did forfeit their lives because of his miscalculation of the UN resolve. Debriefing of captured Iraqi officers noted that Hussein did not believe the United States had the will to suffer casualties to protect oil in a remote area of the world. He was wrong. But more importantly, the threats by the United States and the UN had been perceived as idle, and the diplomatic efforts were misleading. Even an embargo had little impact on the internally based economy of Iraq. In fact, U.S. State Department figures indicate that embargoes are effective only about 30 percent of the time they are employed. That is why technological sanctions provide a necessary, viable option between threats and embargoes, and war.

The circumstances in Bechistan represented the classic conundrum. It was far away, in a section of the world not well understood by most Americans. The geopolitical situation was complex and the relationship and importance to American national interests vague. The U.S. military forces in the immediate area were exclusively based at sea, and reinforcement would be very difficult. In certain Washington-based think tanks and among some senior members of key congressional committees, the true significance of the problem was both noted and understood. Though controversial, it was imperative that regional stability be maintained. A threat to the territorial integrity of western Pakistan or, more specifically, the resources of the area, could not be tolerated.

The first step was to employ diplomatic pressure, directly and indirectly. The leaders of Bechistan seemed to ignore these efforts and continued to move troops closer to their eastern border. The United States issued stern warnings about the consequences of a military threat in the area. These too were ignored. The Bechistan leaders reasoned that the United States would not risk a direct confrontation. At best they had strained relations with all of the nations in the area. Even with increased tension, it was unlikely that Pakistan would allow the United States to land forces in the country or operate from its soil. Therefore, armed intervention by the United States or other Western powers was not deemed a feasible alternative by Bechistan.

For the United States, time would be needed to move both troops and matériel if armed intervention was to be initiated. Actually, intervention was a very unattractive alternative. However, a clear and unequivocal message needed to be sent. It was determined that technological sanctions were appropriate, and the orders were given to proceed. Simultaneously, orders were sent to move a second carrier group from the Pacific into the Indian Ocean. If ground action was required, two carrier groups would provide the minimum force necessary to accomplish the initial phases of the mission. While it was almost unimaginable, America still held a nuclear card. Given the current delicate nuclear imbalance in that region, such weapons could only be used to save American lives if everything went to hell. Even with substantial loss of U.S. troops, crossing the nuclear threshold would be a very difficult decision.

Immediately, intelligence activities for the region were beefed up. Long prior to this devolution of the situation, U.S. intelligence agencies had been provided with extensive, new requirements taskings, the documents the intelligence community used for targeting assignments. The new taskings were significantly different from those of the Cold War days. While the CIA had always provided broad analysis of every country's economic, industrial, and agricultural capabilities, greatly increased emphasis had been placed on infrastructure issues. As best as could be done, data were gathered about each country. Complex models and simulations were developed and run to determine the probable responses to a variety of courses of action. The faults of ethnocentricity were known. Therefore, care had been taken to run the simulations with actions taken from perspectives of people from that geographic area, not how Americans would respond. With some difficulty, even responses considered to be irrational were programmed. While not every possibility could be covered, at least some experience was gained in dealing with "wild cards"—unanticipated reactions. Points of vulnerability had been predetermined. While susceptibility varied, in a global information age all but the most primitive societies had weak spots in information systems.

More recently, as tensions began to mount, businessmen with CIA affiliations visited several cities in Bechistan. There, they surreptitiously deposited telephone devices and forwarded the calls from local phone numbers. These phones were tied to a secret uplink to communications satellites. By monitoring these phones remotely, intelligence analysts would be able to determine the level of success achieved when the telecommunications system was attacked.

Obtaining feedback was a critical issue with several of the non-lethal weapons systems used in technological sanctions. It was necessary to know how effective the attacks were. Traditional intelligence-gathering tools, such as overhead photographic and eavesdropping satellites, were not adequate for the mission. It was essential to report in some detail the information required to make sound decisions about whether or not the objectives were accomplished. Therefore, additional sensors and reporting systems were designed and installed before the onset of hostilities. Still others were deployed after the borders were closed to U.S. travelers. These sensor systems would be capable of monitoring and reporting on a variety of infrastructure functions. Telecommunications was just the beginning.

At the appointed hour, all external communications were cut. For ground systems, this meant simply turning off the connections. Satellite communications were a more difficult problem. As with most nations, communication via satellite was accomplished by rented space on a commercial space-based system. Channels were not permanently assigned, but rather a computer directed each incoming call to the next available line. Taking out an entire satellite was unthinkable since too many countries, including the United States, relied on these systems. Therefore, the attack would have to be subtle and sophisticated so as to not interfere with legitimate users. In this case a code was installed that recognized signals incoming from Bechistan. Rather than retransmitting the signal, the call was blocked.

Signals sent via microwave transmitters were jammed from aircraft lying off the coast. Developed as an extension of our electronic warfare capability, these powerful transmitters could disrupt signals for many kilometers. As part of the intelligence preparation, the operating frequencies of both phone and commercial radio and television stations were gathered. This facilitated tailoring the attack to those specific frequencies. Farther inland, some communications would be possible. This was not a total blackout, only sanctions. Also, communications by fixed landlines would get through, especially the signals traveling on fiber-optic cables.

Immediately prior to the onset of the telecommunication degradation, an execution code was transmitted to all computers that were functioning at the time. That code initiated a massive computer virus attack. The viruses had been previously installed at various times. In a few sensitive systems, the viruses had been installed during the loading of software before delivery. This occurred long before hostilities were contemplated. Known as Trojan Horses, they had lain dormant for years, awaiting a discrete activation code. They were totally transparent and would not be found by any traditional virus-checker program, nor even by skilled computer programmers. Additional viruses and worms were implanted at later times. Some were downloaded with routine executable codes. Others were added during normal maintenance operations conducted from offshore locations. Not all programs were infected. In fact, certain programs, such as those normally employed in health care, were specifically protected from the attack. This was to preclude any unintentional secondary fatalities to innocent civilians.

The collective effect of the computer attack was particularly devastating to Bechistan's financial and industrial organizations. Due to the social structure of Bechistan, the average citizen had little day-to-day dealings with computer technology. In fact, many operated on a barter economy and would be almost impervious to these sanctions. After all, the sanctions were designed to impact the leadership, not the common citizens, who generally had little involvement or influence in national decision-making.

The next round of attacks was launched against the limited ground transportation system of Bechistan. Two technologies were employed. The most effective would degrade the tires on trucks and cars. The other would make high-grade fuel unusable. Since access to large amounts of fuel was not a problem for Bechistan, contamination of the entire supply would not be feasible. Rather, fuels used in the few high-performance engines would be eliminated. In addition to ground transport, aircraft would fall into this category.

As in much of the Third World, the road networks in Bechistan were constrained in many areas. In several mountainous areas, there was a single major artery, often restricted by ravines, bridges, and tunnels. These would make ideal locations in which to get the chemicals into the tires. The weapon of choice was the catalytic depolymerization agent delivered via air-transported caltrops that were scattered generously along constricted areas of the main roads. Emplaced at night, the caltrops were difficult to spot; thus many tires were impregnated with the agent long before the effects began to be noticed. However, by the next afternoon a great number of vehicles were sitting alongside the road, their delaminated tires lying shredded in the wake. None of the tires had blown out; they had merely disassembled as they went down the road. Each had traveled between fifty and one hundred miles since being stuck. Since there was no distinct pattern to the damage that had been inflicted, coupled with the lack of phone communications, it made it hard to realize that an attack had taken place. Even then, finding the location where the agent had been applied would be very difficult for the Bechistan authorities. During that period of imprecise information, more and more vehicles would become incapacitated.

In Bechistan, as in most places, the tires were not easy to replace. The impact on goods and services would mount in the civilian sector. Further, the ability of the Bechistan military to move supplies was also impaired. Their plans to put additional troops in forward positions, and to establish large caches of ammunition and supplies, were dashed. It became clear to the military leaders that if they chose to occupy the mineral-rich territory, they would risk being caught short when defending critical supplies. The senior Army commanders had been educated at foreign military schools and colleges. They were well aware that while young soldiers brag of tactical prowess, professionals knew wars were won by logistics. This insidious attack against the transportation infrastructure would seriously impact their ability to invade and, more importantly, to sustain a force once Pakistan retaliated.

In Washington, at the National Photographic Interpretation Center, known in intelligence circles as NPIC, photo interpreters examined the constant stream of pictures being downloaded from the parade of satellites observing the region. They also received clear photos from stealthy unattended aerial vehicles parked in a nearly invisible mode high above designated target locations. Within a day of the attack against the tires, they had noticed trucks abandoned at the roadsides. There were also increasing numbers of vehicles that remained stationary in auto parks and supply depots. This gave a clear indication that the attack had been successful in destroying Bechistan's ability both to support the population and to mount aggression against Pakistan.

Elsewhere, National Security Agency listening posts monitored the communications traffic from Bechistan, what little there was of it. The leadership was posturing defiantly for the outside world. Internally, a state akin to chaos was developing in the upper echelons of government. Those few civilians with influence in the government were quickly becoming disgruntled. They could make money only when they had free access to world markets. The technological sanctions played havoc with their livelihoods, and their patience wore thin. Although they had asked their alliance partners to intercede, the resulting effort was only half-hearted. India and China would lose a considerable amount if the situation was allowed to escalate to further violence. Realizing that they were being systematically cut off and their military capability seriously impaired, the government of Bechistan pulled its forces back and sent word to the UN that it had no plans to enter the disputed territories.

Though fictionalized, this scenario provides but one of the possible applications of technological sanctions. Naturally, each situation would vary based on the relevant circumstances. There are many different non-lethal weapons that might be brought to bear in these situations. The effectiveness of technological sanctions cannot be guaranteed, but they offer another option between diplomatic sanctions and open conflict. By some estimates, diplomatic sanctions are effective about one-third of the time.[1] Having additional steps could improve those odds.[2]

Technological sanctions could be employed either overtly or covertly. By prior announcement of the intent to the world, it formally puts the enemy on notice that everyone is watching for their response. Of course, that may cause the opposition to dig in and become more belligerent. Therefore, under certain circumstances, it may be advantageous to conduct a strike against a limited infrastructure target. Done properly, by selecting targets with limited visibility—say, an air-defense communication system—only a few people would be aware that the attack had occurred. No direct harm would be caused, and the enemy leadership, in effect, would get a private warning without being exposed to public ridicule from their internal supporters or like-minded bystanders.

There is the possibility that such actions could be misinterpreted. An adversary could take them as an act of war and thus they might precipitate conflict in the near term. It is also possible that the adversary perceives that it has no effective means to counter this threat and decides it is to its advantage to escalate the level of confrontation. How an enemy would respond is part of the evaluation process when determining the courses of action available to high-level decision makers. Before technological sanctions are employed, adequate force should be available to engage and defeat any escalation to violent conflict. Again, technological sanctions offer an option between embargoes and war. They demonstrate that the United States and its allies have the *capability, intent,* and, most importantly, the *will* to take action.[3] If the warning is not taken, then the enemy's ability to prosecute prolonged conflict is already degraded.

Even under the best of circumstances, preventative measures may not be enough. In the next chapter we will explore the possibilities of ratcheting up to another level of strategic application of non-lethal weapons.

14

STRATEGIC PARALYSIS

The best military operation is to attack
strategically, meaning to use unusual tactics.
—*Zhang Yu (Sung Dynasty)*

Without warning, the precision-guided bombs destroyed the headquarters building and surrounding targets. This sudden attack struck fear into the heart of one of the most notorious state sponsors of terrorism, Mohmmar Qadaffi, who had badly misjudged American resolve.[1] Operation El Dorado Canyon, the 1986 air raid on Tripoli attacking both his headquarters and, more importantly, his home, sent a very strong personal message.

A week before, on 5 April, a bomb was detonated at a West German discotheque. More than eighty off-duty soldiers were wounded, and the trail led back to Libya. Qadaffi had guessed that the United States would not take any extreme actions. He was wrong! In retaliation for Qadaffi's role in the bombing, President Reagan had ordered the deadly strike. Hitting late at night on 14 April were twenty-four F-111F Aardvarks, five EF-111A Ravans, fifteen A-6 U.S. Navy Intruders, plus A-7s and F-18s, launched from the USS *America* and the USS *Coral Sea.*[2] With mortiferous accuracy they decimated four of the five assigned targets. Only the port of Tripoli escaped with limited damage.

Beyond the physical damage to Libya, Qadaffi experienced intense psychological trauma. Already nervous about his personal safety after operation El Dorado Canyon, he habitually changed locations every few hours. He no longer had the luxury of spending more than one night in a single bed, for fear that another strike might come. Based on his nefarious past, Qadaffi had reason for trepidation.

From his revolutionary beginnings at the age of sixteen, he had learned to trust no one. In 1969, he preempted other, more-senior military officers and led a successful coup that ousted King Idris I of Libya. Qadaffi, then only twenty-seven years old, expressed desires of being another Nassar, referring to the famed Egyptian leader, and believed he should become the rightful ruler of the Arab world. Ready to support Islamic terrorist movements, he allowed training bases to be established in Libya, provided equipment and travel documents to them. As early as 1972, he had financed the Black September movement and supported their infamous massacre at the Munich Olympics. Even other Arabs were not safe as Qadaffi reportedly engaged the Venezuelan-born terrorist Ilich Ramirez-Sanchez, then better known as Carlos the Jackal, to kidnap oil ministers from Iran and Saudi Arabia.[3] Assisted by Qadaffi, Carlos claimed credit for eighty-three assassinations over two decades before he was finally captured by the French. Qadaffi's support of the Abu Nidal organization led to shootings in London, attacks on El Al and Trans World Airlines in Rome and Vienna, and the hijacking of an Egyptian airliner. He also made training bases available to the Irish Republican Army, the Italian Red Army, and other terrorist groups. Indeed, while befriending sundry outcasts, Mohmmar Qadaffi spawned many enemies.[4]

For nearly a decade following El Dorado Canyon, little was heard directly from this petty tyrant. In 1989, there was a brief confrontation over the Gulf of Sidra. U.S. Navy Tomcats quickly splashed two Libyan fighters, ending the problem temporarily. Intelligence officials believed that the 14 April air raid had convinced Qadaffi that he would be singled out for retribution if firm evidence could tie him personally to terrorist acts. It took a long time before Libyan agents could be associated with the downing of Pan Am Flight 103 over Lockerbie, Scotland. Instead of an attack, UN Security Council Resolution 731 was initiated, demanding that Qadaffi hand over two indicted intelligence agents for trial. His failure to do so resulted in imposition of sanctions that embargoed Libya's civil aviation. His continued refusal brought heightened sanctions, including freezing assets and oil technology in November 1993.[5] Demonstrating his ability to reach out from quarantine, Qadaffi continued to intimidate Libyan expatriates. In November 1995, he orchestrated the brutal murder of an outspoken Libyan dissident in London.

By early 1996, things had taken a much more ominous tone. German intelligence services supplied other Western nations with hard information about a major chemical weapons facility. These reports were believed to be quite accurate, as the blueprints were provided by the German and Austrian companies that had supplied personnel and equipment for the project.

The engineered subterranean cavern, located at Tarhunah, about forty miles southeast of Tripoli, was reported to cover about six square miles. Painstakingly carved out of solid rock, it was claimed that the blast-proof bunker already contained more than 100 tons of chemical agents and could hold much more.[6] Satellite photos of this complex raised concern about the potential for increased, state-sponsored terrorism using weapons of mass destruction. Qadaffi's assertion that Tarhunah was part of an irrigation project was regarded as nonsense.

Bolstered by the success of other terrorist attacks—such as the truck bombing that killed nineteen Americans at their barracks in Dhahran, Saudi Arabia—he was prepared to escalate the scale of provocative action. The threat to regional stability was great enough to cause President Clinton to restate a position of national emergency. In his letter to Congress, the President stated that "Libya's actions and policies pose a continuing unusual and extraordinary threat to the national security and vital foreign interests of the United States."[7]

Despite international warnings, Qadaffi employed his intelligence organization, known as Istikhbaral al Askariya, to carry out more daring and provocative terrorist missions. There were several assassinations in Europe, and even two in Washington, D.C. The press took notice and were beginning to question the security precautions taken by Western nations against terrorist attacks and their ability to deter them. In late summer, Qadaffi crossed the line. NATO's Advisory Group on Aerospace Research and Development (AGARD) was holding a meeting at Pratica di Mare, an Italian Air Force Base located a few miles south of Roma. As was the custom, the Americana Hotel was designated as a common place to sojourn. Located in E.U.R., a suburb on the southern end of Roma, and close to the Laurentina Metro station on the B Line, it accommodated both the bus trips to the air base and nightly visits to Piazza Novona, Piazza Popolo, the Spanish Steps, or the other congregating sites in the center of Roma. The hotel provided accommodations for representatives from eight NATO countries, including the United States, France, Germany, the United Kingdom, and others.

At about 3 A.M. billowing smoke rose from a storage room on the first floor. While smoke detectors sounded the alarm, the fumes quickly filled the halls and stairwells. The guests, shocked into a semiawakened state, stumbled down the stairs, their egress hindered by increasingly dense smoke. Most of those on the first two floors made it outside. Guests coming from the upper floors began to fall unconscious along the escape routes. They were not overcome by smoke inhalation. It was the poison gas—sarin—that had been incorporated into the flammable mixture that produced the fatal consequences.

Coughing and wheezing, the lucky ones managed to slip into the cool night air. They stood there, somewhat dazed and confused by the events unfolding before them. Within minutes, the fire engines began arriving. Of more than 150 registered hotel guests, there were only twenty or so people milling outside the hotel as the first trucks pulled up and began laying their hoses. Then, as the firemen rushed toward the burning building and ten minutes after the fire had been noticed, a Volkswagen van parked in the lot immediately behind the hotel exploded with such force that the foundation of the building shook. The truck bomb had been planted to catch survivors and rescuers alike. Another purpose of the truck bomb was to mask the presence of the chemical agent.

By dawn, the fire was extinguished, and the body count was still rising. In the end, 137 people were confirmed dead, including sixteen Italian firefighters. Another thirty-three people were hospitalized in serious condition, and three were totally unaccounted for. The AGARD study group was decimated. Only two study members had made it out of the building. Of these two, one died in the blast and the other was injured by flying debris. A combination of forensic science and communications intercepts located the responsible agency. The trail led back to Libya.

If the terrorist problem wasn't enough provocation, intelligence reports indicated that Iran had recently shipped to Libya new, long-range, antiship missiles that Iran had

The handheld Maxi Beam 3 million-candlepower light capable of dazzling at a considerable distance. *(Courtesy of John B. Alexander)*

The light produced from this 12-gauge Starflash tactical shotgun shell produces an effective diversion at up to thirty meters. *(Courtesy of John B. Alexander)*

Penetration demonstration with ballistics gelatin. Note that the beanbag penetrated two inches the 9mm SXT round entered fourteen inches into the target. The spiraling pattern of the bullet is easily discernible. *(Courtesy of John B. Alexander)*

Lieutenant Sid Heal, Los Angeles Sheriff's Office, fires the 37mm launcher. *(Courtesy of John B. Alexander)*

Close-up of retractable spikes. *(Courtesy of PMG Manufacturing)*

Police officer with retractable spikes. (*Courtesy of PMG Manufacturing)*

Prototype, lightweight, self-camouflaging caltrop designed by Sid Heal. *(Courtesy of Sid Heal)*

U.S. soldiers train with non-lethal munitions in Europe for peacekeeping in Bosnia, November 1997. *(Courtesy of MSG Bud Schiff—U.S. Army)*

Watermelon penetrated by beanbag round. *(Courtesy of Robert Knight)*

Watermelon struck by a foam baton. One baton is embedded within; another has dented the melon. *(Courtesy of Robert Knight)*

Rubber stingballs barely penetrate the gelatin block. *(Courtesy of Robert Knight)*

The damage done by a beanbag round *(left)* compared to a 9mm slug *(right)*. *(Courtesy of Robert Knight)*

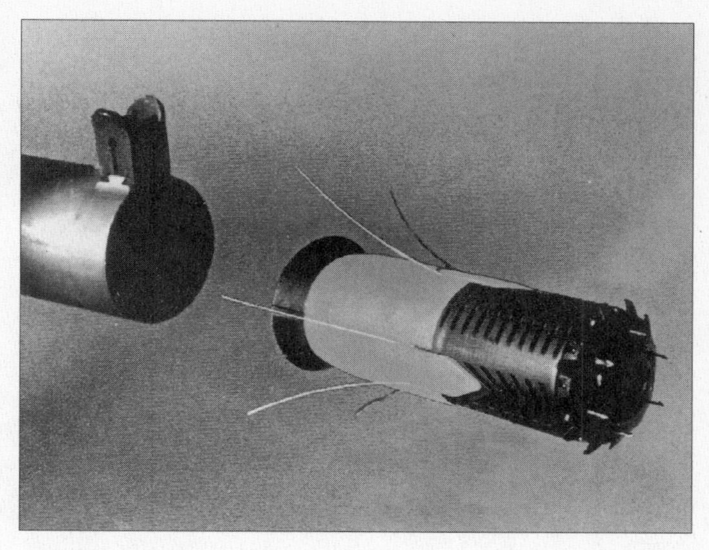

Sticky Shocker 37mm electrical stun grenade as it leaves the barrel. Note the protruding electrodes. *(Courtesy of Jaycor)*

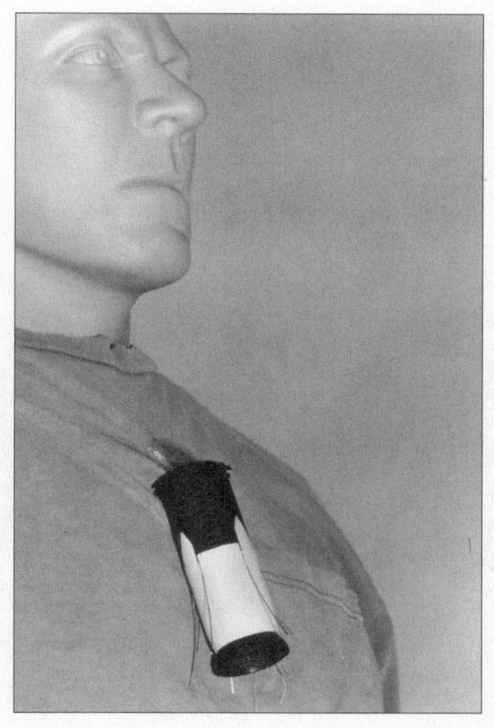

Sticky shocker stuck on mannequin with adhesive. *(Courtesy of Jaycor)*

Truck-mounted Acoustic Blaster. A planar array of multiple acoustic pulsed sources. *(Courtesy of PRiMEX Physics International)*

U.S. Marines test the prototype directional acoustic weapon developed by SARA for the U.S. Army Armaments Research, Development and Engineering Command. *(Courtesy of U.S. Army Armaments Command)*

(a, b) Shoulder-fired net systems that can be used to capture violent or disorderly persons. *(Courtesy of NET GUN)*

(a,b,c) The anti-vehicle Silver Shroud consists of very fine aluminum foil that can wrap up a car or armored vehicle so that the occupants cannot see. *(Courtesy of Foster-Miller Inc.)*

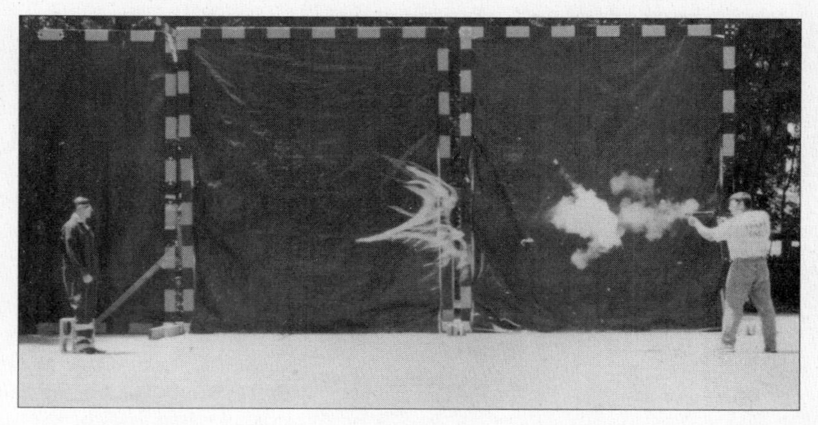

This expanding net fired from a handgun covers a large area. *(Courtesy of Foster-Miller Inc.)*

The Battlefield Optical Surveillance System (BOSS) laser mounted on a Hummer. BOSS blue-green laser being test fired at Kirkland Air Force Base. (*Courtesy of U.S. Air Force Phillips Laboratory)*

Then Lieutenant General Zinni *(left)* observes a demonstration of sticky foam conducted for *60 Minutes* and CBS television. (*Courtesy of U.S. Marine Corps)*

Training with sticky foam on board a ship headed for Somalia. *(Courtesy of U.S. Marine Corps)*

Sticky foam training at Camp Pendleton, California. *(Courtesy of U.S. Marine Corps)*

Marines train with 40mm non-lethal rounds on board a ship en route to Somalia. *Courtesy of U.S. Marine Corps)*

Marines in crowd control training en route to Somalia. *(Courtesy of U.S. Marine Corps)*

Marines train with M-203 weapons on a range run by the Miami Police Department outside Port-au-Prince, Haiti, March 1997. *(Courtesy of U.S. Marine Corps)*

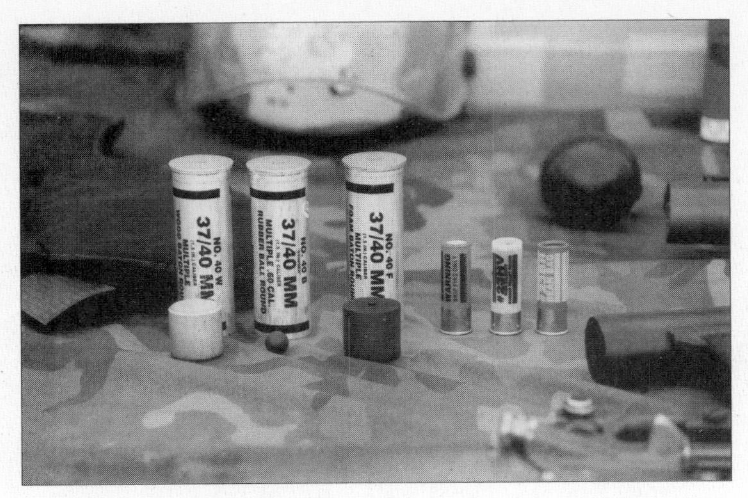

Examples of the non-lethal munitions available to the Marines in Haiti. *(Courtesy of U.S. Marine Corps)*

Prototype sticky foam man-portable dispenser. *(Courtesy of Sandia National Laboratories)*

Prototype counter-sensor laser weapon was designed to look like a conventional M-203. Note that the backpack contains batteries for power. *(Courtesy of Los Alamos National Laboratory)*

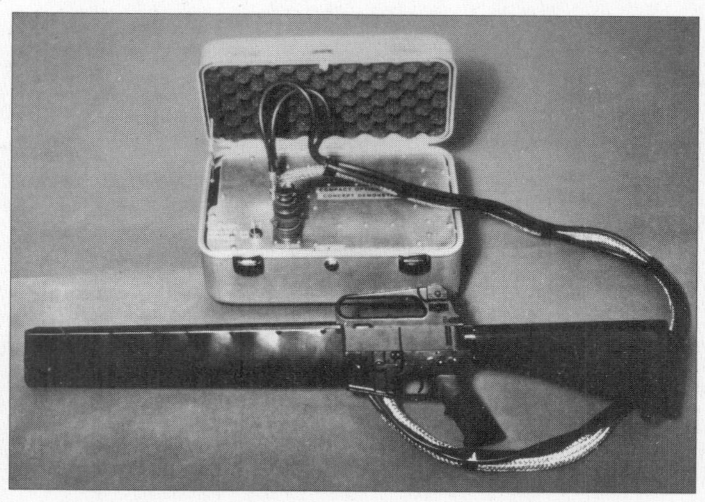

Portable counter-sensor lasers. *(Courtesy of Los Alamos National Laboratory)*

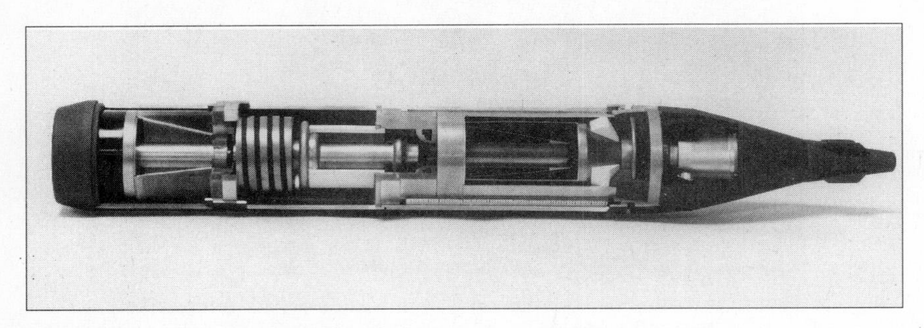

Model of a 155mm RF round. *(Courtesy of Los Alamos National Laboratory)*

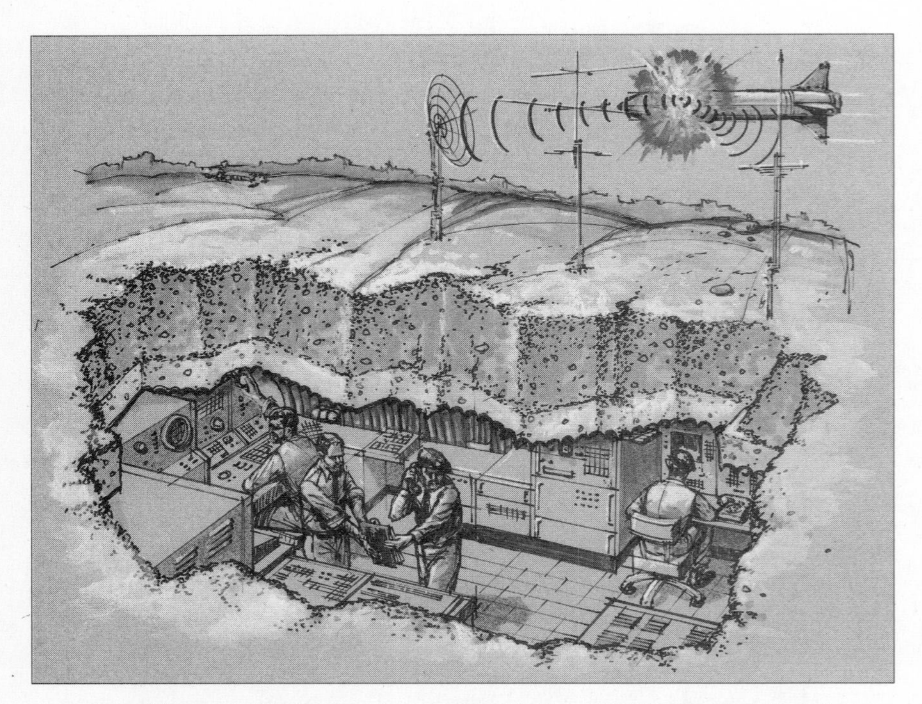

Drawing of how an RF weapon in a cruise missile might be employed against an underground command post. The energy is taken in through the exposed antennae. (*Courtesy of Los Alamos National Laboratory*)

received from China. When Iran first obtained these missiles from China, Secretary of Defense Cohen had warned them that the threat they presented to the Persian Gulf was taken very seriously.[8] *Now these very same missiles could threaten shipping in the Mediterranean. That was unacceptable! The decision was made to force the removal of the missiles and to strike at the terrorist support infrastructure. While not specified in writing, it was agreed that the current head of state in Libya needed to be eliminated.*

Plans for this eventuality had been drawn up years before and had been constantly kept updated. There was only one major decision: How much physical damage would be acceptable in world opinion? While many in the U.S. military pressed for severe punishment, the Allies, though outraged by the recent actions, urged moderation. They reasoned that if sufficient force could be placed on Libya without injuring or killing large numbers of people, they would remove Qadaffi themselves. American officials argued strongly that a similar decision had been made about Saddam—and it didn't work. The Europeans, on the other hand, had other motivations. Pragmatically, they were more strapped for oil. The less damage inflicted during attacks, the more quickly they could reestablish the much-desired flow from Libyan fields. Additionally, even during the years of the embargo, many multinational companies had maintained fairly close ties to the Libyans. They believed they could convince the local populace to take care of the leadership problem, but only if the military was emasculated.

The principles for the attack had been laid out years before by Colonel John Warden, one of the brightest and most innovative officers the U.S. Air Force had produced in decades. Warden formulated the Five Rings Model, which focused pressure on enemy leadership and supporting infrastructure rather than fighting the fielded forces first.[9] *Warden was instrumental in proving his model when the concepts were applied to Desert Storm. Now the model was ideally suited for creating strategic paralysis, which is defined as "attacking or threatening national-level targets that most directly support the enemy's war-making efforts and will to continue the conflict."*[10] *The viable combination of air power and non-lethal weapons had also been described in a study at Air University.*[11]

This operation, now named Sahara Brawl, would be phased to escalate to whatever level of force was necessary to achieve the objectives. That could include a ground invasion, if necessary. An additional U.S. carrier group was dispatched to the Mediterranean, while NATO units began to mobilize to support the invasion force. Immediately, news media around the world began receiving information about Libya's actions. Both the current bombing and potential threat to all nations of southern Europe were emphasized. Daily, press releases were made. If there was no current news, then the releases focused on the long history of Libya's support of terrorist movement.

Meanwhile, pressure in Northern Ireland netted three IRA members who had trained at one of the Libyans' bases. The next day, they were seen on television describing their experiences in the desert of North Africa. The interview was given by them in trade for light sentences by British magistrates. There were also unrelenting stories about the horrors of chemical and biological warfare. Every story mentioned the clandestine facilities located in Libya. Perception management was in full swing.

Taking no time to mobilize, information warriors struck immediately. Scanning for every computer operating in Libya, they started injecting viruses to cause the systems to crash. Having planned for this occasion, the IW troops knew each of the offshore maintenance routines that could be called up for checking and repair of computerized industrial machinery. Once located, they were also crashed. Before long, there were no remaining computer links between Libya and the rest of the world. While Libyan intelligence believed their secret net to Iran was operational and secure, the reality was that it had been completely coopted. The information being traded never went to or from Iran. Instead, it was being orchestrated from Kelly Air Force Base in San Antonio, Texas.

As soon as the naval forces were in position to enforce a no-fly zone, the UN announced the sanction. Late the first night, flying nap-of-the-earth, a few Libyan fighters tried to slip out of the limited airfields, presumably headed for the relative safety of Iran. As they had in El Dorado Canyon, E-2C Hawkeye early-warning aircraft constantly lurked over the area. The look-down radars of the Hawkeye on station easily spotted the evading aircraft. The flight patterns were handed off to waiting NATO fighters, who either eliminated them immediately or escorted them to Egypt, where they would be held captive. UN air supremacy was undeniably demonstrated and would not be challenged for the duration of the operation.

Within the first week, Special Forces and SEAL teams were routinely penetrating Libya and conducting reconnaissance missions. They were specifically targeting the petroleum production and refining facilities both on the coast and farther inland. Since all exports were embargoed, the storage containers would soon fill to capacity. Highest on the target list was the Ras Lanuf Refinery at Ghout El Shaal, which had a capacity to refine 220,000 barrels per day before the 1995 upgrade. Also on the list were the Zaria, Tobruk, and Brega refineries, in that order.

Based on the extremely strong desire to bring Libya to its knees, the decision was made to authorize biocidal-resistant, antipetroleum biological agents into the fuel reserves. It was an ideal special operations mission. As only small amounts of inoculant were necessary to contaminate large holding areas, they could be introduced at any point in the petroleum system. The lay of the land allowed the Libyans to pump oil directly from the wells to dockside refineries. That meant long pipelines that were difficult to guard. The special operations teams could insert the bioagent at points where security was almost nonexistent. The chances for direct confrontation leading to U.S. captives was minimal.

It would be about six weeks before the Sahara Brawl *invasion force was massed. The plan called for both an over-the-shore invasion from the sea and a ground attack from Egypt. Pressure would be incrementally increased—but those would be large leaps. Libya was on full alert and established its armed forces at about 240,000 troops. That number included about a third of the males fit for military service. But of these, a large number were untrained men who had been quickly conscripted. In other wars, such inexperienced military units were called cannon fodder. Predominantly a land force, they were equipped with aging Soviet weaponry—the same kind that had not fared well during Desert Storm. They also had some French, Italian, Czechoslovakian, and Chinese equipment. The outcome of Sahara Brawl, should it come to a land war, would never be in doubt.*

The major cities are all spread out along the coast, which stretches for 1,770 kilometers. The paved-road network was limited to about 27,000 kilometers, not much for a country of 1,759,540 square kilometers—about the size of Alaska.[12] *The limited roads and the geographic constraints imposed by the desert sands would make countertire technology highly desirable. Early in the campaign, caltrop submunitions, filled with catalytic depolymerization agents, were scattered along the major roads every night. While some were discovered and removed, each batch successfully punctured some truck tires and continued to take a toll on the dwindling and irreplaceable supply of tires.*

Thirty-five days into the operation, it was time to put out the lights—literally. That night, air strikes were made against every major power plant in Libya. A few of the more isolated facilities were destroyed by hard bombs. Those close to populated areas were targeted with special weapons containing conductive filaments and foams containing aggressive agents.[13] *The objective was to cause extensive circuit shortages with materials that were not easily removed. The newly upgraded facilities at Tripoli South, Benghazi North, and the largest, a 480-megawatt gas-turbine facility at Khoms, were nearly completely knocked out.*[14] *Further attacks would continue to reinforce the drastic reduction in commercially available power.*

Telecommunications were hit with equal vigor. All commercial television stations were taken off the air. In some cases, HARM antiradiation missiles struck the transmitters and physically destroyed them. When appropriate, cruise missiles with EMP warheads were flown adjacent to the target and detonated, frying the sensitive internal electronics. What little radio communications remained were effectively jammed with conventional electronic warfare equipment. There were a few fiber-optic links between the major cities. A lesson had been learned in Desert Storm about the difficulty of slicing underground cables with a "big ax from the sky." This time, special operations forces had prepared these telecommunications links days in advance. Digging down to the cables in isolated areas, they had planted remote-control charges at critical locations. Along with the charges, a sophisticated technique was employed so that reflectometers could not locate the positions of the fractures. As soon as the air raids began, the ground links were severed. Quickly, the head was being separated from the body.

The chemical facility at Tarhunah represented a special problem for Sahara Brawl planners. Actually, it was an ideal target for a redesigned, low-yield, B-61 Mod 11 nuclear weapon.[15] *The concept had been made public in 1997, but employment was effectively killed through political pressure from antinuclear weapons groups. Deeply buried, hard targets such as this were exactly what that bomb could have been employed to eliminate. Instead, it would have to be taken out the hard way.*

Three options included: penetrating conventional bombs that, if accurately delivered, would cause the underground passages and airways to collapse; self-sealing, non-lethal weapons; and, finally, a ground assault. Option three was very undesirable, and option one difficult to execute. Careful study of the aerial photographs revealed the location of most, if not all, of the ventilation openings, as well as passages for personnel to come and go. Thus, option two was chosen.

The first attack would take place with precision-guided munitions carrying small microspheres of aggressive agents. These would be dispersed near the ventilation openings, each of which had a strong draft to force the air into the underground facility. The microspheres would be rapidly inhaled into the filters, there to break and begin to eat through them. The destruction of the filters would render the chemical plant unsafe.

Sensor-triggered alarms should go off, sounding an alert, and the concerned people would evacuate their work areas. The second attack would contain stabilized foams that would block all air going into the factory. Since those trapped inside during phase two would likely die, messages would be dropped and broadcast shortly before the second attack, warning the occupants to depart immediately.

However, there was another target that also lay near Tarhunah. In late 1997, an extensive tunnel system consisting of 2,000 miles of pipe thirteen feet in diameter was brought to the world's attention. Similar to one created by Saddam Hussein near Kuwait, it crossed Libya from Egypt to Tunisia and headed south, almost reaching Sudan and Chad. Though claimed to be a water pipeline, the size and configuration were indicative of another, more insidious purpose. It was large enough to allow substantial numbers of Libyans to move close to the borders without being seen. The coincidental proximity to the chem-bio facility did not go unnoticed. Given Qadaffi's relations with his neighbors, it was more likely that the tunnels would be used for military purposes rather than for moving water.

Based on information derived from ground-penetrating radar, the tunnel system was targeted to be cut in several places. These strikes would be lethal for anyone unfortunate enough to be inside. The deep-penetrating bombs would rupture the pipe and disperse a fuel air explosive that would follow the contours of the pipe. When the proper chemical ratio was achieved, it would detonate, causing overpressure, killing anyone inside, and leading to collapse of large segments of the tunnel system. Planners wanted to ensure that there would not be any Libyan units emerging in unexpected places.

What had begun as a perception management operation became a full-fledged psychological operation. One of the techniques used was the air delivery of thousands of radios capable of receiving only one station—the one the UN forces controlled. It was the only method by which civilians could receive information from the outside world. Effectively controlling all information to the public, they were constantly told that the objectives were limited but nonnegotiable. It was in their best interest to have the current government surrender or, failing that, to remove it. If these conditions were not met in short order, a full ground invasion would follow. Casualties, so far very limited in number, would necessarily rise dramatically.

The Libyans were reminded of the humiliating defeat that was handed to Saddam Hussein when he tried to resist the UN forces. Saddam, it was pointed out, had a much larger and more experienced armored force, one that had been crushed in less than one hundred hours of ground combat. Resistance was futile. Lacking adequate defenses, the Libyan defeat would be even quicker than Iraq's. By quickly surrendering, they could preserve their cities from total annihilation. The forces of the UN had no quarrel with the Libyan people, only with the misguided leadership that had brought dishonor to the country and might provoke absolute carnage.

The psychological operation was sound. It was based on the results of similar efforts in Desert Storm. The numbers were impressive. In post-testing of Iraqi prisoners, it was learned that 98 percent of the POWs had been exposed to the leaflets, and 80 percent actually believed the message.[16] *A large number of soldiers decided to surrender because of the impact of those leaflets. Based on the Desert Storm results, the PsyOps campaign in support of Sahara Brawl had the potential to save thousands of lives. If not, there was always phase two. However, phase one, using non-lethal weapons in combination with selected precision-guided munitions while limiting collateral casualties, had severely degraded the petroleum, transportation, electrical, and communications systems across Libya. Strategic paralysis was accomplished. Should the ground invasion be necessary, the ability of the Libyan armed forces to resist had been severely impaired.*

This scenario is one application of strategic paralysis employing non-lethal weapons. It is more likely that such high-end conflict would be executed with a combination of lethal and non-lethal systems. Isolated military targets would probably be destroyed with precision bombs and missiles. However, non-lethal weapons provide the capability to strike targets while minimizing the potential for collateral casualties.

Strategic paralysis represents an extreme example for application of non-lethal weapons. On the other end of the spectrum are domestic law enforcement operations. Let us now consider how some of these weapons might assist police confronted with a difficult hostage situation.

15

HOSTAGE/BARRICADE SITUATIONS

Are you gonna come and kill me?
—*A Branch Davidian child to FBI hostage negotiator Jim Cavanaugh*

"In Vietnam, I was an Air Force helicopter pilot that supported SOG," the unidentified voice on the radio began. The program was *Coast-to-Coast,* hosted by popular Las Vegas–based radio personality Art Bell. Bell rules the late night radio waves across the country. Carried by more than 400 stations and with 10 to 12 million listeners, he specializes in controversial topics, including conspiracy theories. The subject this night, Gulf War Syndrome, brought about a lively debate.

While I was somewhat interested in the Gulf War Syndrome controversy, the mere mention of SOG caused immediate arousal within me at a most visceral level. Studies and Observations Group, or SOG, was the most highly decorated unit in Vietnam. Comprised of Special Forces soldiers and supported by indigenous troops, they pulled off some of the most daring and dangerous escapades of the entire war.[1] Operating deep behind enemy lines, they frequently had to be extracted under extremely heavy fire from the North Vietnamese troops. Too often they didn't make it out.

The voice, I hoped, was going to provide information about the conduct of these top-secret missions and the heroism of the reconnaissance teams on the ground, and the air crews that supported them. The caller then described what he called *Sawdust.* "We would fly over the troops in contact and dump *Sawdust* on them. It would knock everybody unconscious immediately. Then we would go in, pick up our guys, and shoot the bad guys." Instantly, the caller's credibility plummeted to zero.

After returning from Vietnam, I had worked for Jack Singlaub, who, as a colonel, had been the SOG commander for many of the critical years. Outspoken and forthright, Major General Singlaub retired from the Army after a public disagreement with his commander in chief, President Jimmy Carter. To verify my contention, I called Major General Singlaub at his northern Virginia home. While he had admitted on national television that SOG had employed non-lethal chemical agents on these cross-border operations, he assured me that no universal incapacitating agent was available.[2] The program, CNN's *Impact,* twisted his statements to sound as if the introduction of temporary incapacitating chemical agents were part of a nefarious, knowingly illegal plot.[3] The reality was that Singlaub was desperately trying to find a way to reduce collateral casualties among innocent civilians who wandered into SOG patrols. Deep in enemy territory, the alternative was to kill them.

For reasons previously discussed, there is no known substance that can quickly—and safely—incapacitate all people. Variations in health conditions and body chemistry just don't allow it. Unfortunately, many gullible people heard the caller on the Art Bell show and uncritically accepted his unfounded claims.[4] Indeed, a tool like *sawdust,* more often described as *magic dust,* would be ideal for hostage situations. It was needed at Waco, Ruby Ridge, the Philadelphia MOVE eviction, and a host of other sensitive situations, ones in which too many lives have been lost. It didn't exist then. It doesn't exist now. Since we don't have it, nor will we anytime soon, we must use the tools that are available. Consider this scenario, which incorporates non-lethal technology that actually is available.

It was the one of the worst situations imaginable. Shortly before noon, a lone gunman had entered Wilson Elementary School located in a middle-class suburban area of Denver, Colorado. Shots were fired; casualties, if any, were unknown. Unopposed, the gunman had quickly captured the startled principal, threatened the administrative aide with his assault rifle, and smashed the face of an assistant who moved too slowly. With his rifle pressed tightly under the principal's chin, the gunman roughly escorted him to the third-grade classroom down the hall.

Alerted by the noise of gunshots, the teachers closest to the administrative offices realized that something was dreadfully wrong. There were no plans or drills for rampaging gunmen. Acting instinctively, teachers moved to protect their students. Those who could directed their wards to exit via windows and to get away as fast as possible. The idea of establishing rendezvous points might have been nice, but the reality was that they were running for their lives without time to plan or think. Two men making a delivery from a van parked near the rear of the school heard the commotion and darted out of sight into the maintenance room.

One alert teacher grabbed her bag, pulled out a cell phone, and dialed 911. Her voice trembling, she informed the police dispatcher that an emergency was in progress at the school. Few details were available. She mentioned hearing shots, but couldn't confirm where they came from or how many people were involved. Despite the dispatcher's plea, the teacher folded the phone, turned, and dove through the first-floor window, landing in a recently trimmed bush.

Almost immediately the police radios crackled with alert notifications and assignments. The SWAT team assembly was activated. Of course reporters, who often monitor police frequencies, received information simultaneously. Without hesitation, radio and television stations prepared to break into programs in progress. For parents who would learn that their children were in mortal danger, the abrupt and uncaring notification would remain indelibly imprinted for life. For everyone listening, the nature of the call evoked a sense of extreme urgency and foreboding. The mayor, police in neighboring communities, and even hospitals were notified of impending disaster.

A distraught madman with an AK-47 assault rifle in a grade school: What could be worse? The answer came almost immediately. The gunman wasn't alone, nor was he a mentally impaired lunatic. As the turbulence began to subside, the two delivery men, now wearing masks and carrying guns, reemerged into the nearly vacant hallway. This action had been planned in detail, and was being carried out by three members of the God's Frontiersmen, a previously unknown militia group. Denver had been chosen because this was the city in which Tim McVeigh had been tried, convicted, and sentenced to death. As far as the Frontiersmen were concerned, McVeigh was set up in the Oklahoma bombing. They actually believed the rumors that the DEA had blown up the Alfred P. Murrah Federal Building in order to blame militia groups, gain sympathy from the American public in general, and use the incident as an excuse to wipe out the groups.

Word of the gunmen spread throughout the school. With the exception of two classes, all of the other students and teachers somehow managed to reach the outside of the building nearly unscathed. In the haste to escape, there were a number of children and teachers who collided with desks, doors, and each other. The bumps and bruises sustained were relatively minor. Recounting tales of close calls, the teachers congratulated themselves on getting so many out of the building in such a short time.

In fact, their easy departure was part of the plan. The Frontiersmen knew they could only handle a limited number of hostages. They purposely chose a third- and fourth-grade class to remain. Younger students were likely to cry a lot and place a tremendous burden with "potty requests" and the like. Older students, having watched more TV and movies, might try to do something heroic and stupid that could result in unwanted injuries.

The police quickly cordoned off the entire area. A mobile command post was brought in, helicopters stationed overhead, and assembly areas for reinforcements established. A hostage negotiator called the school but received no answer. Using a bullhorn, the negotiator asked to speak with the kidnappers. Simultaneously, a lieutenant, trained in public relations, was arranging to address the growing number of impatient media representatives. As with military operations, perception management would be very important.

Obtaining accurate intelligence was an initial concern. While the police suspected there was more than one person involved, they still didn't know how many there actually were, or what their motive was. After about an hour of constant dialing, someone in the school answered the phone. Providing few clues to their identity or purpose, the hostage taker informed the police that this was a revolutionary act and the people involved were prepared to give their lives for the cause. If the police made any attempt to rush the building, many of the children would die needlessly. It was noted in the conversation that the gunman seemed to know both police tactics and the current situation. The reality was that part of the planning had included renting an apartment with a clear view of the school. In addition to the three men inside, two young women were observing the police operations from the apartment and relaying information via cell phones—all stolen, of course.

To consolidate the kidnappers' efforts, the students and their teachers were moved to the gymnasium. This location offered many advantages. Locker rooms, with no external exits, provided bathroom facilities. The only windows were high enough that direct observation was avoided. Also, one gunman could watch all of the hostages while the others were free to roam. And since the gym was located next to the cafeteria, food would be accessible with minimum movement.

From a police perspective, the situation was very difficult. While a terrorist would occasionally be visible long enough for a clean head shot by hidden snipers, never were all three visible at one time. Further, the close proximity of the children made even precision, high-power rifle fire undesirable. A stalemate had begun. The SWAT team attempted to move closer to the school. Each time the team maneuvered, a call came from the terrorists stating that they were aware of their tactics and threatening to kill hostages. After several attempts, the field commander ordered the teams back to their observation posts. However, through careful analysis of the taped conversations, the police came to the conclusion that the terrorists had accomplices outside of the school. Exactly where they were located could not be determined, but only a few positions could provide good observation of both the school and the activities of the police.

The hours dragged on without comment from the terrorists. They had come prepared to wait it out and wanted to demonstrate that they were prepared for a long haul. Unfortunately, the parents of the students were not. Neither were the media. By the six o'clock news, questions were being raised about the effectiveness and competency of the police force. National attention was already being focused on this small Denver suburb. The people of Colorado really did not want another situation like the JonBenet Ramsey case, in which the Boulder police were severely criticized for their handling of that case.

At about 7:00 P.M., the lead terrorist contacted the police. His demands included release of a number of militia people serving prison sentences around the country. Further, they wanted McVeigh's death sentence commuted to life imprisonment, guaranteed in writing by the President. They also demanded that the government release the real files on Waco and Ruby Ridge, not the doctored versions already available to the public. Finally, they wanted to be flown out of the United States to an unspecified location. The government had forty-eight hours to meet those demands. The caller indicated that no further negotiations would be held. However, in case the police thought about sneaking into the school in an uninhabited area, they should know that booby traps had been placed at numerous locations. Detonation of one of those devices would be taken as meaning an attack was underway, and drastic action would follow. The police now knew this was not going to end quickly.

The night would provide some cover for the police. Even if the terrorists had night-vision equipment, they would not have the same degree of coverage they enjoyed during the day. Under the relative protection of darkness, the police dropped small acoustic sensors from passing helicopters. For now, the media were inadvertently providing cover for the delivery operation. From the listening devices, the police were able to determine the general location of the terrorists and hostages. Periodically, noise was heard in differ-

ent areas of the school. The terrorists were patrolling the halls. The pattern of movement provided clues about where the booby traps might be placed. Over time, large areas could be eliminated as being mined. Once inside the building, the hallways appeared to be clear.

The next morning, a sick girl was released unannounced through the front door. It was a major break. With a temperature of 102 degrees, the terrorists did not want one sick child contaminating the others. It was a tough decision for them, but they opted to send her out. Though only nine years old, she was able to confirm the number of terrorists and their primary locations. Careful questioning resulted in substantial information about the actions of all members of the group. Children observe a lot of details, more than they are given credit for. When first asked about a situation, they will usually provide an accurate response. However, repetitive questioning may lead them to confabulate to meet the perceived wishes of the adult asking the questions.

The day passed with little contact. The police were exasperated at their inability to resolve the situation. The media began to complain about being left in the dark and, lacking details, began to focus attention on the competency of the police. Live-action cameras were constantly beaming up to orbiting satellites. In addition to Denver stations, CNN and the networks were carrying news about the hostage standoff. Of course they were being broadcast back down as well. Both the terrorists inside and their accomplices outside were being continuously updated in real time.

The few times officers moved in close proximity to the school, the Frontiersmen let it be known that they were under observation. Once, when an officer came a little too close, a shot was fired. It dug up the dry dirt five feet in front of the officer and produced a dust trail for all to see. Clearly, it was a warning and not intended to provoke a gunfight. The inability to determine the location of the external support people was extremely frustrating to the law enforcement officers, who were now augmented with state and federal agents. With young children involved, the situation had to end, and end quickly with minimum risk to the hostages.

That evening, at the request of the terrorists, fresh milk was delivered to the front door of the school. It was the break officials had been waiting for with cautious optimism. Embedded in the milk cases were acoustic sensors. Once in the school, these would provide better information about the situation as it progressed. Two boys from the fourth grade retrieved the milk and delivered it inside the building. The transmissions lasted about ten minutes. Then the cases and the wax containers were thrown back out the front door. The Frontiersmen had studied Waco. They knew about how tiny transmitters were slipped into the Davidian compound and wouldn't be fooled by the same trick.[5] In fact, sensors had been embedded, anticipating they might be found. The real surprise was not discovered.

The decision was made to end the siege that night. Special equipment had been brought into the area to support the rescue operation. Shortly after eleven o'clock, quiet generators began creating mist. Taking advantage of the gentle breeze, the mist would engulf the adjacent buildings and obstruct the vision of the external group. That would allow the rescuers more freedom of movement. Black, unlit helicopters, with ducted fan engines that greatly suppressed the sound of the rotor blades, appeared at about eleven-thirty and hovered about one hundred feet above the school. That altitude improved the chances for an undetected insertion by preventing rotor wash and minimizing any sound that could give them away. Silently, two men, covered head to toe in black, rappelled onto the roof. Equipment was deftly lowered, and in less than a minute the helicopter was gone. Even the news crews had not spotted the action. Of course it helped that, at that minute, the chief of police was some distance away giving a briefing for the media.

With padded footgear, the SWAT team members moved to their designated point of first entry. It was to be through a skylight several rooms away from the gymnasium. With care, they placed the large suction cup on the pane of glass and made sure the vacuum seal was tight. Next, superacid, carried in a Teflon-lined container, was applied to the glass near the outer perimeter of the window. The hole had to be large enough to allow men to drop in without fear of getting cut or creating a sound of broken glass. The superacid completely ate through the glass in a matter of minutes. Using the suction device, the pane was lifted onto the roof. Next, low-light-level cameras were lowered into the room to allow continuous monitoring of the point of entry.

After preparing the entry point, the team moved to the area of the roof above the gym. Here they quietly drilled a series of tiny holes through the ceiling and inserted fiber-optic lenses. These fed back to shielded TV transmitters operated on special frequencies that would not normally be detected by civilian scanners. The five cameras, operating with wide-angle lenses, would now permit field commanders a decent view of the area of operation. The first pictures confirmed that the mild sedative that had been injected into the milk had done its job. The children were all sound asleep. The raiders would not have to worry about children being in unanticipated locations once the operation began. The view also confirmed that the terrorists were monitoring CNN. They knew that intelligence agencies around the world practiced this procedure.

In the meantime, the communications van had taped the eleven o'clock news broadcast on CNN. During the day, special arrangements had been made with the local television cable provider. When alerted, they were to replay the eleven o'clock version of the news. That would prevent any inadvertent broadcast of the actual operation that might tip off either the terrorists inside or their supporters on the outside.

At 1:45 A.M., two black helicopters returned, and six members of the Denver SWAT slid down nylon ropes and alighted ever so softly on the roof. There they joined the two members who were already on the roof. They were assigned in four two-man teams. Moving to the gaping hole in the skylight, they checked the darkened room below. Stealthily, they descended to the floor and prepared to engage anyone who entered. They knew that at about 2:00 A.M., one of the terrorists usually checked the building. The intent was to take him alone, leaving only two to eliminate in the gym.

With the arrival of the helicopters, low-power RF jammers were turned on. They were similar to those used in electronic warfare to block air-defense radars, but just strong enough to block all cellular phone transmissions in the immediate area. Even if the spotters noticed the impending action, they could not reach their compatriots by phone. Should they attempt to run to the building, snipers were positioned to remove that threat. Effectively, the school was enshrouded with an electrical and visual obscurant curtain. Television via cable was now running broadcasts from the earlier time slot.

Two members of the SWAT Team One waited in ambush at a hallway intersection.

Two others remained hidden in a room that the terrorist would have to pass. Once he had passed this point, the team members would cover a section of the floor with extremely slippery oil. In the unlikely event that he evaded the first team, the terrorist was effectively prevented from returning to his companions. Between that point and the gym, Team Two activated canisters that released rapidly expanding foam to fill the hallway. The purpose of the foam was to dampen any sound made when the terrorist on patrol was taken out.

The Frontiersmen had already become a bit complacent. They had roamed the building for two days and felt confident that either their booby traps or external "eyes and ears" would provide warning if an action was imminent. Also, they believed the police would wait for a few days before attempting a daring rescue. It would prove to be a terminal mistake.

As the roving guard came to the intersection, he heard a quiet thud as a compressed air-gun launched a dart. Before he could open his mouth, the fast-acting neurotoxin had paralyzed his motor functions. One SWAT member grabbed his gun, while the other caught the falling body. Unlike the ambiguity of preparing a universal incapacitant, the approximate weights of the terrorists were known, so an effective, but nonlethal, dose could be calculated. He would awake in a few minutes, bound, gagged, and arrested.

Cautiously, SWAT Teams Two, Three, and Four moved toward the gym. There was no doubt they could overpower the two remaining terrorists. The trick was to do that without injury to the young hostages. They had ruled out using explosives or flashbangs as a distraction. Since the children were sleeping, they did not want any of them to react unexpectedly, possibly getting in the line of fire. Instead, they had installed a small speaker in the ceiling. An indistinguishable noise would be played, loud enough to attract attention of those awake but not to startle the sleeping children. Waiting outside the gym doors, the SWAT team members identified the two terrorists. They were located on opposite sides of the gym. On command, the speaker came on. As the terrorists instinctively looked up, the SWAT teams burst into the room. The one terrorist nearest to the door recoiled as he was hit by the air-powered stun gun. Though he had reached for his assault rifle, his body went into involuntary convulsions before he touched it. The voltage was kept on until the first team member reached him.

The second terrorist was more problematic. Beirg farther away, there was no nonlethal weapon that could instantaneously affect him—except light. Three red lasers formed a triangle pattern, two on his chest and one on his left eye.[6] *He had to make an immediate decision. The slightest move, even a muscle tremor, and he would die instantly. Without hesitation, he chose not to be a martyr. The highly skilled marksmen held their fire, although they would have been fully justified in terminating any of the terrorists on the spot. The children, many still sleeping, did not even know the ordeal was over until police officers and school officials entered and woke them.*

This is only one of many possible hostage and barricade scenarios that police will face in the future. Most will not be as complex. More likely, they will be one of the following: the lone gunman holding a family member or ex-girlfriend hostage; the mentally deranged individual; bank robbers caught in the act and who take employees or patrons hostage and attempt to negotiate their way out. With time to get into position and bring specialized equipment to the scene, non-lethal alternatives make sense.

On the other hand, criminals may come prepared for a shootout similar to the bank robbery at a North Hollywood branch of Bank of America in Los Angeles on 28 February 1997. Wearing full body armor, the two criminals believed themselves to be invincible and acted accordingly. For such a situation, systems that take away vision or otherwise incapacitate the individual may be useful. However, a high-powered elephant gun is probably the best solution.

Part IV

THE ISSUES

Non-lethal weapons are not universally embraced. They will not be totally accepted until they have proven their worth in real-world hostile situations. For those who have been in difficult situations, such as those faced in Somalia, Haiti, Bosnia, and Panama, the need for non-lethal weapons is clear. As we have seen, similar situations—ones where lethal force has limited value—are likely to be more prevalent in the future.

However, there are many valid concerns about non-lethal weapons. A debate is just beginning. It will go on for several years to come. We must address issues of law, ethics, technology, policy, and psychology. Soldiers must come to have confidence in these new systems. Therefore, non-lethal weapons must prove to be reliable and effective. They should be fielded only with adequate lethal force as backup, and only after troops are thoroughly trained with the new systems. Premature distribution of non-lethal systems is one of the biggest dangers this fledgling field faces.

Part IV addresses most of the known issues relating to non-lethal weapons, their development and employment. The final chapter, "Winning," will address one of the most sensitive issues about future conflict: What constitutes winning? The issue strikes at the heart of our value system. Many will find the notion of redefining winning controversial and maybe even distasteful. Let the debate begin here.

16

LIMITATIONS

Assess yourselves and your opponents.
—Ho Yanxi (Sung Dynasty)

One warm summer night in September 1956, as a slick-sleeve private at Fort Chaffee, Arkansas, I was selected to go on guard duty for the first time in my military career. With less than a month on active duty in the Army, my basic training company had not yet officially learned how to shoot. Still, I was assigned to a post that required possession of live ammunition. The Sergeant of the Guard asked me if I knew how to fire the M-1 rifle we had drawn from the arms room. I told him I had done some hunting while I was growing up in La Crosse, Wisconsin. He said, "Okay, Alexander, here is a clip of live ammunition. Whatever you do, don't shoot!"

Forty years later, how far have we come? From personal experience, I can say that giving a young soldier live ammunition and then telling him or her not to shoot puts that soldier in an untenable position. While troops today undoubtedly have more training than I received prior to being issued live rounds, the situation is basically the same—and maybe worse. Now we ask soldiers to make split-second, life-and-death decisions in terribly complex situations. They are given the rules of engagement (ROE) by which to make those decisions. Whatever their choice, soldiers will be judged both by military standards and the court of public opinion as their actions are broadcast around the world by CNN.

In times past, the rules of engagement were pretty straightforward: If the bad guys come, shoot them. In peacetime it was a bit different, but still simple. When one walked a guard post, any intruder would be challenged. Failure to receive a proper response was authorization to use force. Lethal force would be used to protect arms and other critical commodities. The media had little access to incidents in and around military bases. If there was an accidental or questionable incident, in general it was handled internally. From a public perspective, the term *national security* was accepted at face value. Rarely did the news media aggressively question military authorities about internal incidents. Even when dealing with public episodes, it was presumed that the military was acting under proper authority. The technology of the day did not allow for live coverage of unfolding events. Rather, information was provided to the public a day or more later. In retrospect, we frequently ask about how certain experiments, such as exposure to radiation, came to be approved and carried out. In talking to the participants, the conversation almost invariably begins with the statement, "It was a different time."

Social consciousness was also different. The world was divided into "good guys" and "bad guys." Winning was interpreted as physical defeat of the enemy—who were by definition "bad guys." Therefore, the safety and well-being of those relegated to "bad guy" status were irrelevant. In fact, to make war more tolerable, extensive measures were taken to dehumanize the enemy.

During World War II, American propaganda posters depicted Japanese and German soldiers with ratlike features. Racial manifestations were also employed, such as in Vietnam where the enemy were referred to as "slopes," "dinks," and "gooks." Similar treatment was true for people in a position to be associated with the "bad guys," be they active supporters or just there through a happenstance of birth. Issues such as "collateral casualties," while sometimes intentionally avoided, frequently were not. Dehumanization of the adversary made it easier to rationalize additional killing.

Early in the development of non-lethal concepts, I presented an argument about how Americans prided themselves on their high moral values. In discussion with retired Air Force Major General Cecil Powell, the general confronted me with another reality about U.S. behavior in war.[1] The reality was that we have not always acted in the most humanitarian manner possible. And this was not just individual acts of barbarism. He cited acts supported by conscious, high-level decision making. For example, during World War II we intentionally fire-bombed the cities of Dresden, Germany, and Tokyo, Japan. While there were military targets in both areas, civilians in those cities bore the real brunt of the attack. In fact, Tokyo, with its buildings constructed of extremely flammable material, had been so devastated by the fires that it was taken off the target list for a site of the first atomic-bomb attack. Targeteers were concerned that the Japanese military would not realize the true extent of the awesome power of this new weapon. The prior conflagration would mask the new damage.[2]

The dawn of the Information Age had dramatic impact in shaping all future conflict. The reporting on the war in Vietnam was the harbinger that brought about the distinct change. For the first time, television would bring the conflict into American living rooms on a daily basis, and war would never be the same. From then on, the time between events and broadcast was shortened. By Desert Storm, that time had been shortened to real-time television coverage of U.S. missiles landing in Iraq. Due to broadcast constraints with allied forces, we frequently saw the missiles hit their targets before getting pictures of the launch. Pilots flying attacks against Baghdad could race back to their bases in Saudi Arabia and watch their bombs strike on CNN.

The coverage of world events by CNN is so pervasive that almost every key office and operations center around the globe keeps a television set constantly tuned to that channel. This is true from the Pentagon to the CIA to the White House itself. Almost every nation conducts a similar vigil. At the slightest sensing of trouble, CNN cameras are there, recording every move. Their presence at violent encounters is so common it has become known in the military as the "CNN effect." The impact is immeasurable, particularly on peace support operations. No longer can soldiers act recklessly without fear of being instantly reported. There is no rewriting of history by the winner. In real time, or near real time, people around the world know what has happened.

This means that use of deadly force will be viewed and adjudicated in the court of public opinion. If deemed inappropriate, it can have a devastating effect on the support rendered to the troops in the area. However, these same visual forces can also be used to muster support when troops do act appropriately. Many groups have learned how to manipulate the media. Events such as the small group of protesters that caused the *Harlan County* to sail away from Haiti can easily be staged. The media, always looking for the breaking story, are often ready to oblige them—warranted or not.

A major concern is that the presence of TV cameras may cause soldiers to hesitate in their decision to use deadly force. As both soldiers and law enforcement officials know, such hesitation can be fatal to them or to their comrades. This is where the need for effective and proven non-lethal weapons can make a major difference. The availability of non-lethal weapons, supported by lethal systems, can be more readily employed in marginal situations. Once the ROE for the use of force have been met, troops would be free to use those weapons. Given the reduced likelihood of fatalities or serious injuries, the reason for hesitation should be eliminated. Again, no system will be perfect, so use without provocation would be unacceptable.

In the development of non-lethal weapons, one consideration must be how such weapons will be represented by the media. There may be serious tradeoffs between effectiveness and visual impact. For instance, we all remember the pictures of Rodney King being beaten by the Los Angeles police officers. One of the weapons used by the police was an electrical stun device called a Taser. Repeatedly, King was subjected to shocks. While most people agree that police officers should have the tools necessary to subdue a criminal suspect, they do not approve of punishment using these kinds of devices. Seeing a person flopping about due to electrical shock can be just the picture that loses support for the police on that department.

There is another piece to the Rodney King incident, one the public never was allowed to see. When it was shown on television, the news media intentionally deleted the first eighteen seconds of the video. While it does not vindicate the police for use of excessive force, it clearly indicates that it was King, not the police who initiated the attack.

"Did you see those Marines come ashore in Somalia?" one retired marine general officer asked me. Though I protested that the encounter with the news media had not gone well, he went on. "They came in looking tough and ready. A bit of a show, but the Somalis got the message not to mess with them." In fact, the stated agenda for U.S. forces going into Somalia was "Go in hard and get out quickly."[3] Of course, now we know better. But then one out of two wasn't too bad. Still, there are those who believe that a show of force is sufficient to accomplish such missions. They assume a rational adversary who will fear for their lives if they get into a shooting war with U.S. troops.

This cultural-centric position also assumes that potential adversaries have the same values regarding loss of life—their own and others'. Unfortunately, that is not a realistic view of today's world. There are two driving factors that can make people behave in ways that appear illogical in our eyes. These factors are *culture* and *poverty.*

In many areas of the world, poverty levels have reached the point at which loss of one's life is not a major concern. Given a choice, the individual would probably choose to live. However, given the choice between living in destitution and taking a high-risk chance on obtaining something better, he or she will take the risk. The population growth in most developing nations is so great as to engender despair in a large segment of the populace. With nothing to lose, taking up arms against a technically superior force is easy. Of course, they will choose to be smart about it. In the American tradition of *Roger's Rangers, Mosby's Rangers,* and *Merrill's Marauders,* and the *Studies and Observations Group (SOG),* these adversaries will use unconventional methods. They will quickly learn to wait, watch, and "hit us where we ain't."[4] Instead of being the innovators of new concepts, it is likely that we will be taught that being on the receiving end of such tactics is not much fun. In fact, these unconventional tactics can be very frustrating to the defender who finds the options for using force extremely limited. Non-lethal responses will be necessary to fight terrorists in close proximity to civilians.

Cultural values play an important role. In Vietnam, the normally gregarious and friendly GIs frequently became the victims of a grenade attack launched by children. Our natural tendency for trusting and helping the young, old, and infirm has been identified as a weakness by others. Meanwhile, our adversaries are prone to impressing others to make attacks against us. The risk of such attacks is high and the price for getting caught even higher. Nonetheless, in almost every conflict in which we have become engaged, they continue to occur.

The attacker may be naive, dedicated, or coerced through fear of pain or loss of a loved one. The reason matters little. However, the cold, calculating ones who plan these operations have no compunction about sending innocent people to their deaths to further their cause. In wars past, as well as in current conflicts, merciless foes have herded civilians through mine fields, used them as human shields when initiating frontal attacks, and as willing hostages to provide cover for snipers. The decision makers are prepared to lose hundreds, even thousands, of lives to achieve their objectives. Properly orchestrated, they can now gain strategic advantages by exploiting "the CNN effect."

Another countercultural issue is the concept of self-induced martyrdom. In most technically advanced countries, preservation of one's life is assumed to be the highest priority, giving one's life to save another is an act of selfless sacrifice, but to commit suicide intentionally to further a cause is almost unthinkable. Yet there are societies in which suicidal acts committed in support of a holy cause are not only acceptable, but are considered honorable. In the Middle East, some groups espouse the idea that giving one's life in such an attack ensures martyrdom and an instant transition to heaven. Thus, terrorist planners have no difficulty recruiting young men to carry out missions of no return.

The concept is not unique to the Middle East. During the final phase of World War II, the Japanese developed the kamikaze corps. The kamikaze, or *Divine Wind,* derived their heritage from a typhoon in 1281 that destroyed the invasion fleet of Kublai Khan, thus saving the island nation. Not seen as a futile gesture from the tradition of the Samurai, these missions were the honorable alternative to surrender.

In that tradition, during World War II many young, very inexperienced pilots volunteered to be strapped into planes loaded with high explosives. Then, provided with only enough fuel for a one-way trip, they were sent into the Pacific Ocean in the Battle of Leyte Gulf to dive-bomb the advancing American fleet. Their effectiveness is attested to with over thirty ships sunk and hundreds of others damaged.

While they could not turn the tide of the war, those pilots were important in influencing actions taken by the Allied Forces. The attacks signaled the determination with which Japan would fight and an indication of the casualties awaiting any invading troops. It was partially based on such determined resistance that the decision was made to drop the atomic bomb on a major military target rather than presenting a demonstration.

Clearly, we have seen that willingness to accept casualties on the part of an adversary places troops with different belief systems at a significant disadvantage. Suicide bombers in the Middle East and elsewhere have demonstrated their willingness to die for their cause. Accommodating martyrdom tactics—that is, killing the assailants—is actually playing into their long-term strategy. While it is almost impossible to stop someone from blowing himself up, non-lethal weapons can minimize the number choosing martyrdom. You get no points if you are captured and incarcerated for a long time.

In addition to situations in which troops are facing an elusive enemy, PSOs frequently encounter crowds that are fundamentally friendly but still want to extract something from them. Examples include food distribution to semi-starving people or protection in "safe areas."

In the humanitarian role, it is still necessary to protect both the troops and the aid workers. The people being assisted are frequently quite desperate. Situations have arisen in which these people come to believe that food supplies, or other critical resources, are inadequate to support the assembled population. Riots break out as people scramble to make sure they get their share before the supply runs out. Without thinking about the consequences, they attack the supplies and anyone who gets in their way. Lethal weapons would only exacerbate the situation and will never pass the CNN test.

In Bosnia, UN forces faced another problem. They were assigned to ensure the integrity of designated "safe havens." With limited troops available, it became necessary to admit defeat and extract our forces to areas that could be protected. The people who had assembled in these safe havens became extremely frightened—and with good cause. Since the onset of hostilities, the population had been subjected to unspeakable horrors, thousands summarily—often brutally—executed, woman raped, children tortured, and homes burned, all in the name of ethnic cleansing. They had fled to the UN safe havens in sheer desperation. And now the only thing between them and continued murder and torture, the UN troops, were leaving them to the waiting onslaught. Naturally, they did the only thing possible. They got as close to the UN troops as they could and tried to block their retreat.

In such a situation, when diplomatic actions have failed and the forces must be extracted, non-lethal weapons provide the only viable alternative. This is a situation in which there are no good choices. Still, the countries providing troops to PSOs do not place them there to fight for one side or another. Public will is not strong enough to risk casualties to the UN peacekeepers. Therefore, for such intractable situations, non-lethal weapons are an essential force augmentation; not for stopping slaughter, unfortunate as that may be, but just to get our forces out alive.

Difficult situations are not limited to foreign soil. In August 1992 *Hurricane Andrew,* the most devastating natural disaster in American history, slammed ashore and devastated southern Dade County, Florida. With winds so fierce that anemometers broke, Andrew killed thirty-eight people and left a quarter of a million people homeless. When the counting was done, there was an estimated $20 billion in property damage, enough to bankrupt some smaller insurance companies. The public response was immediate, and relief supplies were quickly mustered in Broward and Palm Beach Counties and then transported to the heavily damaged area south of Miami.

My son, Marc Alexander, was one of the deputy sheriffs from Palm Beach County who helped transport food and clothing to Dade County. The first day of the relief effort, they went unarmed and in civilian clothes. To their shock, some aid trucks were hijacked at gunpoint before they could reach the hardest-hit areas. Because of the robberies of convoys that occurred, every day after that, the deputy sheriffs were in uniform and fully armed. Two of my brothers, Donald and George, and their families, lived in the center of Andrew's ferocious path. Both lost their homes—not blown down, but blown away. To protect the meager belongings that were the residue from his destroyed home, my brother Don, a Metro-Dade Fire Department lieutenant, chose to stay there with a loaded shotgun. To ensure that looters didn't make off with the remaining scraps during the night, he slept on the rubble, gun in one hand and a dog tied to his other wrist to alert him if anyone tried to approach. Anything not protected was stolen. The police were incapable of coping with a disaster of this magnitude. In fact, many of them were struggling to find and protect their own families. In only hours the highly structured society that was once Dade County disintegrated into mob rule. It took a week of martial law enforced by the U.S. Army 82nd Airborne Division and Florida National Guard units to restore law and order. During that time, some Guard members operating without ammunition had their weapons stolen by armed gangs.

This is similar to the situation we saw on television when parts of Los Angeles erupted in violence after the first verdict was rendered on Rodney King's assailants. Within a short period of time, the police had lost control of some sectors, fires were raging, and people were being beaten and killed by rioters. Controlling riots is extremely difficult. When limited to lethal force, the task is even more complicated. While you can stop the violence temporarily, the root cause continues to fester, waiting for the next triggering event. Use of lethal force during riot suppression will lead to more violence at a later time. Coupled with the CNN effect, it can actually swing the balance in favor of the rioters in the long run. Alternatives must be sought that facilitate law enforcement agencies regaining control during such outbreaks of violence.

Current lethal systems, with which the military and law enforcement agencies are armed, have definite limitations. They can be used to intimidate or they can be used to kill. We have learned the hard way that a show of force cannot be effective unless the threat to use that force can be substantiated. That leaves troops with the options of threatening or killing—nothing in between. We urgently need to provide military troops and law enforcement officers with options that provide a credible, incremental ability to ratchet up levels of force. This means development of additional non-lethal weapons.

17

STRATEGIC IMPLICATIONS

The Way means inducing people to have the same aim as the leadership.
—*Sun Tzu*

The Media

We have become a society of sound bites and clichés. Driven by thirty minute television news programs and hourly, five-minute interruptions of radio music programs, we demand that the most complex topics be reduced to one-line summaries. The business motto is, "Don't bother me with the details; just give me the bottom line." That bottom line is best described in simple terms such as "Do it or don't do it," "Sign or don't sign," "For or against," and "Good guys or bad guys," each a simple dichotomy. Gray is unknown in this domain. Decisions made in this manner have strategic implications. They can determine whether or not we go to war. These sound bites are used to inform and condition the people in order to gain support for the decision. Those in opposition employ the same methods of communications. It is the media who determine what they will report and transmit those messages to thus molding public opinion.

These simplistic solutions are not relegated to uneducated people making unimportant decisions. At the Kennedy School of Government Program for Senior Executives in National and International Security at Harvard University, Mickey Edwards, a former Republican United States Congressman with sixteen years' experience in the House with service on both the powerful Budget Committee and the Appropriations Committee, shared his insight into how Congress *really* works and how decisions get made on the Hill. In many ways his talk was truly frightening.[1] Congressman Edwards described how he would leave for a vote on a topic he knew little, if anything, about. He stated that as he walked out the door of his office, he would ask his administrative assistant, "What is the issue, and how should I vote?" During his walk, various lobbyists would approach him and advise him how to vote on the bill. If he was still not sure which way to cast his ballot, he would walk into the House chamber and see how others were voting. If he saw those with a similar political philosophy as his voting for the issue, he, too, would vote for it. The converse observation would precipitate a vote against.

From my dealings with senators, congressmen, and their staffs, I believe Edwards is not unique. It is not that they don't care—they do—but they are constantly on information overload. To illustrate the effect of this overload, Tipper Gore once told me that if she were not mentally prepared to see her mother, she might not recognize her if she bumped into her accidentally on the street. While this is a bit of an exaggeration, it reflects the information load politicians have and the constant bombardment they endure on a wide variety of issues. At that time, Al Gore was still a congressman and one of the brightest I encountered. I know the information requirements have gotten worse since, and there are finite limits to what anyone can remember. Too frequently, political decision makers jump to conclusions based on limited data. Getting them to change their minds, particularly if they have made a public statement on the issue, is an extremely difficult task. With a few exceptions, most congressmen do not have access to any special knowledge about the issue on which they vote. *The Washington Post, The New York Times,* CNN, and possibly a major paper from their district are likely to be their principle sources of information.

The reduction to the bottom line is endemic. Large businesses and government agencies must synthesize huge amounts of information. The synthesis process goes on continuously, until final actions are taken. While staff workers will normally do extensive studies about all of the issues, it is usually a one-page executive summary that the CEO or politician sees and acts on. He or she must trust subordinates to have done due diligence on that issue. However, beyond a limited number of well-informed staff members, few in large organizations, including senior executives, are really informed about many important issues.

The background material frequently comes from Lexis, Nexis, or similar information services. Since these files are rarely purged or corrected, incorrect information resides unchallenged forever—even if formally disproved. Since news reporters frequently do not possess the information necessary to discriminate between the good and the bad sources, they too often include erroneous information in their pieces, thus further perpetuating the problem.

Another significant problem in information storage and retrieval systems is the lack of adequate source or origin identification. Information from both very credible and noncredible sources often has equal weight due to virtual anonymity of the real origin. Even worse, if the bad information has a spectacular quality to it, many media outlets will pick up the story. Emotional topics such as abortion, chemical and biological warfare, cold fusion, and, more recently, cloning result in phenomenal amounts of inaccurate reports, all of which are filed without discrimination. Confronted with the sheer volume of dramatic reporting on the topic found in the archives, the reporter is likely to develop a bias not supported by facts.

Headlines exacerbate the problem even more. They are usually not written by the author of the article but by an editor. In fact, the intent of these two people may be very different. The author should want to impart facts about an issue. The headline editor wants to attract attention to the story. Unfortunately, this motivation can generate misleading headlines. Too often, readers scan a newspaper for articles of interest and go no farther than the potentially misleading headline. Then they get the "details at eleven."

Compounding the problem is the news media. Challenged by time and space constraints, often having insufficient credentials in the topic being covered, the reporter must quickly obtain background information, followed by the new data, and convert the mélange into something intelligible to laypeople—all against a deadline.

While some news organizations do attempt in-depth coverage of selected topics, these are usually done in retrospect and garner relatively small audiences. Anyone doubting the tastes and attention of American viewers need only examine the weekly television program ratings. In general, you will find us more interested in *Friends* than any international issue. The only news programs to attract large audiences, such as *60 Minutes, 20/20,* and *Dateline,* tend to have an exposé quality to them. They are closer to entertainment than delivery of facts on complex issues.

Education

Education is no longer just a social issue. Education is a matter of national security! The importance of education in the interrelated fields of economics, the environment, and social processes continues to increase dramatically. In an age of constant data overload, what information is ingested and retained will be critical. In his 1997 State of the Union Address, President Clinton talked extensively about the need to improve education in America. He too noted that education was a matter of national security.[2] Unfortunately, the news media reporting on that event seem to have totally missed that point and concentrated on the costs and social benefits. Our leaders will need to be more prepared than ever before to face the threats of the future.

What prepares senior decision makers for their posts? Their education and experience. Raised on distilled and simplified information, students in our educational systems also require synthesized programs. With rare exceptions, history, geography, and languages are abysmally neglected. This frequently results in a we-versus-they mentality when considering international issues. We make minimal effort to learn about others and often state, "If they want to talk to me, they can learn English." As Americans, we know little about the world beyond U.S. borders and generally couldn't care less about the rest of the globe. Since Americans represent less than 5 percent of the population of the world but use in excess of 25 percent of the expendable resources, these are dangerous beliefs to hold and espouse.

I vividly remember my partner at Command and General Staff College, Major Zafar Abbas, a brilliant armor officer from Pakistan, complaining about news in America. He pointed out that each network's broadcast contained the same information, just played three times. Except for large-scale violence or major natural disasters, little was shown about daily events in the Third World. Zafar stated that he had better access to information about world events in Islamabad than anywhere in the United States.[3]

Since hard sciences and mathematics do not lend themselves to "pabulumization," we find that many American college students shy away from them, particularly at the graduate level. Although American universities rank very high in academic prowess, surveys of attendees at graduate-school level in physics, chemistry, and mathematics show that a very large percentage of the students are foreign. This means that we are exporting the knowledge required to develop future advanced weapons systems. At a minimum we need to insure that we continue to educate Americans in the hard scientific skills required for weapons systems development as well as for countermeasures.

The educational problem starts long before graduate school. Our schools spend too much time telling students *what* to think, and not enough time on *how* to think. Thus, with quiet nods to social promotions through high school level, we have developed one or more generations that are woefully unprepared to understand the complex nature of the world confronting them. The problem is so pervasive that President Clinton has formally addressed the issue.[4] In preparation for a world in which highly skilled workers can expect to have five or more major career changes in their lives, information-acquisition skills and an ability to conceptualize systems will be of paramount interest. For those wanting to survive and be successful in the *Information Age,* learning a single lifelong occupation is a concept quickly passing into oblivion. With globalization of the economy and emergence of developing countries as true competitors, it is imperative that we alter our educational system and prepare students to meet these new realities.

Three significant national security issues arise from our underperforming educational system. First, we will have a limited pool of young men and women from which to recruit. Weapons systems are getting more and more sophisticated. The soldiers, sailors, and airmen who are the weapons operators must be capable of using complex systems. Tanks—the relics of battlefields past, yet still dominant in land warfare of the future—now have sophisticated range-finding and fire-control systems that require the crew members to use basic computer skills. Even "grunts," the infantrymen, have become *digital warriors.* They now hold in their hands Global Positioning System devices that, based on satellite data links, display the unit's exact position. Once these infantrymen owned the darkness through use of state-of-the-art night-vision equipment. Now others have acquired such tools, thus complicating night warfare many-fold. Aircraft, intelligence systems, air-defense weapons, fighting ships, and submarines are constantly becoming more sophisticated.

As we upgrade, so too do potential foes possess increasingly capable systems. Heavy reliance is made on equipment manufactured in countries of the former Soviet Union. Cash poor, they have sold many of their advanced weapons systems to developing nations to support foundering internal economies. In the Cold War era, nations receiving such systems usually had a leash on them. Either the United States or the Soviet Union could control employment, either directly or indirectly through logistics. The result of the recent fire sales is that nations that previously could not have developed sophisticated weapons can now buy them in the burgeoning international arms bazaar. Not to be outdone, the United States also plays a key role in this market. This means that American troops may well have to fight against American-made weapons systems.

The vast increase in complexity of the battlefield requires soldiers who are extremely bright and well educated. In future conflicts, it will not be the low end of the social scale that protects the educated elite. From the top decision makers to the lowest private, seaman, or airman, individual survival will depend on intellectual as well as physical skills.

Non-lethal weapons will not ease this problem. In dangerous scenarios, troops on the ground frequently will be required to make difficult choices. Always having lethal force available, it will be the individual soldier, in the midst of an unclear situation, who makes the decision which weapon to use. It may be a young man or woman under twenty years of age, likely frightened, who sets the course of destiny. Education and training in tactics, overall mission objectives, and an understanding of why they are there could be critical. A wrong decision by a junior officer or noncommissioned officer in a peace support operation may have international consequences.

Senior commanders must determine the rules of engagement and communicate them clearly and unambiguously. Despite satellite communications systems that allow the Pentagon, or even the White House, to talk to front-line commanders, mission exigencies will push critical decision-making to lower levels than ever before. Decisions once made by generals will, of necessity, be relegated to field-grade officers, who may be the senior U.S. representative on the ground. This makes it imperative that these officers be schooled not only in tactics, but also have a fundamental understanding of international politics. The training of these officers must come both from changes in military schools and from their civilian education.

The second education factor of strategic importance will be the education level of the civilian population, which ultimately must understand and approve the actions taken by the government. Unfortunately, many do not have the educational background to comprehend current issues of national security, and the issues of the future will be even more complex. With no concept of global spatial relationships, it is difficult to envision interactions occurring at a distance, particularly if we have rarely, if ever, heard of the place. Remember, before 1989, few Americans ever heard of Bosnia-Herzegovina, Chechnya, or Byelorussia, all prominent in today's news. Still fewer could find them on a blank map!

International issues may be closer than you think. Canada, which is ethnically similar to the northern states, is viewed by many U.S. citizens almost as an unannexed territory. On the other hand, Mexico, with more than three times Canada's population, is almost unknown to those not of Hispanic descent. (Tourist areas such as Acapulco and Cozumel don't count.) Yet here is a country with long contiguous borders with ours that has recently experienced major financial instability and an ongoing insurrection in Chiapas, been a major transition point in the illegal drug trade, and has a population-growth problem that is nearly out of control. While we bemoan illegal immigrants and pass racially biased laws to repress their influx, few take the time to understand the underlying causes of this migration. The fact is, there is now a global south-to-north migration that is going to affect the standard of living in all developed nations. Suppression will work only for a limited period of time. We are seeing only the tip of the iceberg in illegal immigration. If we are to take actions, we must first understand the fundamental issues so that we will respond responsibly. That can only occur if we cultivate an educated, socially conscious citizenry.

One of the significant issues interrelating national security and education is foreign aid. An example of our lack of understanding of the importance of international support was seen when President Clinton presented his 1998 Budget proposal. In it, he recommended that we increase the amount of spending on foreign aid for the first time in many years. Since 1992, foreign aid has declined 51 percent. There is a general perception by Americans that we send a large amount of money abroad only to have it squandered by inept or corrupt foreign government officials.

When surveyed, it was found that Americans believed that 15 percent of our American federal budget was being sent overseas. Those surveyed indicated that they thought 5 percent would be reasonable. The reality is that less than 1 percent is spent on foreign aid.[5] This extremely low figure is at a time when regional political instability is endemic, unemployment in technically advanced countries is high, and there is concern about the spread of technical capability and materials to develop weapons of mass destruction. An understanding of those complex factors would lead us logically to conclude that we should be expanding our outreach dramatically. This is not necessarily an altruistic measure but one in the best interest of our national security. The American public needs to be better informed.

Non-lethal weapons provide commanders and politicians with more options in application of force. They are but one of the many pieces necessary for resolution of the thorny problems of the future. At least the decision makers will not be limited to bombing neighbors versus allowing threats to national security to go unchecked due to reluctance to use excessive force. An educated population, one that understands the sophisticated, often Byzantine issues related to national interests, is an absolute necessity.

The third education issue is in weapons development. As weapons systems become more sophisticated, the brainpower to design them must also improve. As we mentioned earlier, many of the students studying hard sciences in our graduate schools are foreign. Often, countries require their students who study in the United States to return and practice in their home nation. That means we are training and exporting a large number of people who are capable of developing very sophisticated weapons.

A conceptual problem already exists within the U.S. military. A by-product of endemic American *cerebralcentrism,* the belief that we are the only smart people in the world, many military analysts and leaders ignore the advances being made by others. Naively, these staff officers think that if the United States makes a discovery, no one else will find it later. European weapons designers often query experts in the field about technologies that U.S. developers have classified because they thought they had found a breakthrough, one not known by anyone else. Physics and chemistry work the same for everyone, and there are many smart people working the same technologies. What our cerebralcentric approach really accomplishes is that, while a few of our weapons developers are aware of the advanced technology, U.S. field commanders are not. Meanwhile, foreign weapons designers are developing these technologies, possibly for use against us. It is our troops whom we place at risk through excessive secrecy.

One technical area that is rapidly being exploited aboard is computer software. Education, as well as economics, are driving factors. Many computer companies have established operations in Bangalore, India. This area has been referred to as the new Silicon Valley, so named for the world-famous center of software development in central California. One major reason for using Indian computer programmers instead of Americans is that their basic education is different. If an American is asked to debug a program, he or she will quickly isolate the problem and fix it. Indian programmers, however, will go to first-order principles and fix *all* of the problems, even those not yet discovered. Of course, modest wages are another attractive feature.

This offshore trend also means that many key component systems are being designed, assembled, or programmed outside the control of U.S. supervision. This is a serious concern to American weapons procurement programs. As sophisticated weapons usually rely on computer programs at several stages in their functioning, it will be important to know where each and every part comes from. With increasing international cooperation in development of common components, it is reasonable to question what the concept "American-made weapons system" really means.

Environmental Issues

Environmental issues will play a greater role in future conflict. Those issues will include the resources that are fought over, such as petroleum and minerals, and the impact of weapons systems on the ecosystem.

With the population mounting, scientists are debating the upper limit, or *carrying capacity,* of the Earth. That number is supposed to represent the population the world's resources can sustain at any given time. Estimates place the number at about 9 billion people, a population we will reach between 2020 and 2030. In fact, when graphed, current population trends are approaching exponential growth. Increased population will undoubtedly lead to increased tension and conflict over available resources.

Great disparity in resource usage now exists between the *"haves"* and the *"have-nots"* around the globe. With recent alterations in weather patterns, large numbers of people face famine each year. Although sufficient food is raised to support the current population level, distribution systems, especially in the developing countries, are quite inadequate. A small percentage of technically developed nations use most of the resources. Populations grow at the same time that developing nations are demanding access to a greater percentage of those resources. This places a rapidly increasing burden on demand. Coupled with a need for immediate cash, nations can make short-term decisions that have long-term, very negative consequences. For instance, many hardwood forests are being cut as a cash crop. Due to the long time it takes for the forest to recover, irreparable damage is sustained.[6]

Deforestation, combined with use of certain technologies demanded by developed nations to maintain, or improve, their standard of living, is creating what many scientists believe is a global warming trend, or the *Greenhouse Effect.* Known culprits, and ones we are reluctant to give up, include chlorofluorocarbons in refrigeration, emissions from engines, electrical power generation, and anything that increases carbon dioxide (CO_2) or methane (CH_4) content in the atmosphere. While international organizations and governments are now developing treaties to minimize the impact of acts deemed to be injurious to the planet, there will be constant tension between expediency and long-term environmental needs. There is a tendency to assume that some scientific breakthrough will occur that will magically cure all the problems we are consciously, and unconsciously, creating. This is unlikely to happen, and the result of procrastination will be increased conflict. We may be faced with decisions about waging war over acquisition or protection of vital resources. Whether or not we are successful at curbing our appetites remains to be seen. If we are not successful, then new force options are in order. Non-lethal weapons capable of protecting natural resources and the infrastructure that develops and distributes them will be a necessary alternative.

Political Considerations

"How long are you willing to stay there?" That was my response when I was first asked about using non-lethal weapons in the former Yugoslavia. While non-lethal weapons may seem attractive for such a scenario, a bigger question must be answered: What is the objective, and what price are you willing to pay? The former Soviet Union has already proven that with sufficient force, internal hostilities can be held to a minimum. In post–World War II, with substantial help from Marshal Tito, Yugoslavia appeared to be a relatively peaceful, well-integrated society—at least to the outside world. In fact, beneath the surface, hostilities, some decades if not centuries in the making, had been festering and once the oppression of the Soviet Union was removed, these hatreds ferociously erupted.

While non-lethal weapons may be useful in quelling or minimizing a conflict, there must be a political determination that the outcome is maintainable. In a case such as we see in Bosnia, use of non-lethal, or lethal weapons for that matter, provides only a temporary solution. If all weapons were somehow magically removed from the area, left to their own devices people would use sticks and stones and, when necessary, bite each other. In other words, in the long run, non-lethal weapons would have little advantage over lethal systems, and the effect would only last as long as a physical presence was maintained. Therefore, before committing troops to any conflict, it is essential to understand clearly and articulate the desired outcomes. Unless directly threatened, such as was the case with NATO, the United States usually is not good at long-term commitment of forces. Among the first questions we hear from Congress when considering employment of troops is, "How soon will they leave?"[7]

At present, the ability of the United States, the UN, and NATO to respond to intrastate problems is extremely limited. In general, there has been a reluctance for such organizations to become involved in affairs internal to any given nation. The "killing fields" of Cambodia stand as mute testament to the disastrous consequences of allowing internal aggression to go unchallenged. Currently, the devolution of nation-states precipitates numerous ethnically based conflicts, which pit groups identified as indigenous to a given geographic area against other, similar groups. Due to those geographic boundaries—which are really artificially imposed—the conflict is defined as internal, and therefore not to be interfered with.

Occasionally, such conflicts pose a sufficiently significant threat that they attract external attention. The March 1998 attack in the Yugoslav province of Kosovo was one of those. Serb forces assaulted the town of Srbica ostensibly to prevent an ethnic Albanian separatist movement from gaining strength. More than 80 people were killed, including several women and children as young as five years of age. By today's standards, the total number killed was not particularly alarming. Rather, it was the potential for spread of violence throughout the region that brought the immediate attention of the international community. Of concern to Americans was the close proximity of U.S. peacekeepers in Macedonia and the problems that could arise if actions involved Greece or Turkey. This relatively minor interaction, left unchecked, could undermine NATO. It is an example of small problems that have strategic importance.

The problem had emerged earlier. In Belgrade in 1989, then-President Bush informed the Yugoslavian leader, Slobodan Milosevic, that excessive force against ethnic Albanians in Kosovo would bring an armed response by America. That policy was reaffirmed by the Clinton administration. Additionally, sanctions on arms buying were imposed by several nations. Missing from the equation are non-lethal weapons options. Basically, we are stuck with bombing Serbian targets, air power demonstrations like those of June 1998, or rhetoric.

While Balkan-type scenarios do not bode well, they point to situations in which non-lethal weapons do offer strategic political advantage. Prior to the onset of hostilities, they can be used to send a strong message that we fully intend to engage the adversary. By attacking infrastructure targets, non-lethal weapons have the ability to make our intentions crystal clear while degrading the enemy's ability to wage long-term war. We cannot prevent an enemy from launching an immediate strike with the weapons already in the field; no system can do that, and even a highly lethal preemptive strike will not eliminate all weapons. However, degradation of the infrastructure offers the potential enemy the option of backing down before an all-out assault is launched by us, one in which their losses will be very high. Non-lethal alternatives make it easier to retreat from the brink of disaster. The infrastructure is repairable, while lives lost cannot be restored. This is what Gen. Edward C. "Shy" Meyer, former U.S. Army Chief of Staff, termed "the Death Barrier,"[8] that is, a threshold that, once crossed, prohibits return to the status quo.

Using non-lethal weapons to send a message does have a potential downside. If the adversary feels trapped or is not prepared to back down, it may choose to initiate offensive actions. Extreme caution must be taken in the political and military analysis of the situation before *any* weapons are employed. As in all situations, sufficient lethal force must be available to respond, should the adversary choose to escalate rather than capitulate.

A third strategic alternative for non-lethal weapons usage is to inhibit mobilization of two external adversaries. These are places in the world where it would be disadvantageous to everyone if a war broke out. India and Pakistan offer one such possibility. Both are declared nuclear states, and have tested nuclear weapons. These strategic weapons are in an unbalanced position due to the third contentious party in the area, China. The potential for conflict in the region is high. Animosities run long and deep. If one party initiates a nuclear strike a devastating exchange of weapons of mass destruction will ensue. Such an exchange would make resolution nearly impossible and would affect all of Asia, if not the world. In past conflicts, resolution has been achieved over time, but rarely without bloodshed. Time compression is now a reality, meaning that things now happen faster than they used to. Forces can be deployed more quickly and, once they are employed, greater damage can occur in a very short time. This time compression, very evident since the Gulf War, works against peaceful solutions. Therefore, there is a need to develop means by which the time for diplomacy can be extended.

The concept entails the use of non-lethal weapons to slow the escalation, even though the United States would have no intention of entering the conflict on either side, especially with ground forces. The targets would be transportation, communications, and other systems that were supporting the mobilization effort. If perceived threats can be reduced, time can be gained during which a diplomatic settlement might be reached. Such a scenario would be high risk, but possibly worthwhile if a nuclear confrontation could be averted.

Some antimatériel non-lethal weapons have the added advantage of minimizing physical damage to the target. This is an important issue, as history has shown that we frequently bear the cost of rebuilding the vanquished. In a situation such as Just Cause in Panama, we had a vested interest in holding down physical damage from the inception of the operation. When employing non-lethal weapons as a warning or to inhibit others from conflict, offers to assist in repairs may be made as further incentives to comply with our desires.

While the benefits from use of non-lethal weapons in peace support operations are acknowledged by almost everyone in military leadership roles, the strategic implications are not. Several senior commanders, including General Anthony Zinni, USMC, commander in chief, U.S. Central Command, and General John Sheehan, USMC, have spoken to the strategic application of non-lethal weapons.[9,10] Both have personally been involved in operations in which non-lethal weapons were a necessary alternative.

The reason non-lethal weapons will become increasingly necessary is the changing nature of conflicts in which the United States will become involved. The role of information in decision-making, our educational system, and the environment will all greatly affect when and how we employ our military forces in the future. Non-lethal weapons provide alternatives for difficult political situations. Most importantly, before any amount of force is used, it is essential that the long-term consequences be thoroughly thought out. To remain a viable power, senior military and political leaders must begin to address these complex issues.

18

IN OPPOSITION

A new scientific truth does not triumph by convincing its opponents . . . but rather because its opponents eventually die.

—*Max Planck*

"Non-lethal weapons are not a panacea," began Deputy Secretary of Defense John Deutch. He was restating the obvious as he addressed our Council on Foreign Relations study group. For three hours we had discussed the virtues of non-lethal weapons, so the comment was appropriate to ensure that we were not out of balance. In all of the sessions the group held, no one *ever* suggested that non-lethal weapons would solve all, or even most, of the problems imposed by the necessity for use of force. The study group was composed of several people with extensive experience at a senior policy-making level who would not be easily enamored by technology or hyperbole. We had certainly not suggested that non-lethal weapons replace lethal systems, nor was the group of one mind about all of the ramifications regarding their development and employment. Rather, we became convinced that these options were required to accommodate some of the complex problems of national security the United States would face in the near future.[1]

In fact, there are many serious and complex issues concerning non-lethal weapons that must be addressed. Those issues encompass such diverse topic areas as policy, ethics, technology, psychology, and economics. The legal aspects are extremely important and have been relegated to a separate chapter. As this book comes off the press, the non-lethal weapons debate is really just beginning in earnest. In the next few years there will be many more studies, articles, and papers written about non-lethal weapons. It is not reasonable for these weapons to be acclaimed as a *blinding flash of the obvious.* Rather, technology and, more importantly, real-world situations will increase the pressure for acceptance of these weapons systems. Solutions to the controversy will not come easily.

The first policy issue that is always mentioned in non-lethal weapons debates is known as the "slippery slope" argument. The rationale goes that if non-lethal weapons are available, they make it easier for policy makers to decide to commit U.S. troops to dangerous situations around the world. Once those forces become engaged in or near hostile activity, they may experience "mission creep" and get more deeply involved than initially intended. Because the first action was authorized, committing to additional responsibilities is viewed as "supporting the troops" instead of making a new policy decision based on the change in facts at that time. Some members of Congress and other political analysts feel that having non-lethal options could make a President more adventuresome. It is argued that if the reality of death and destruction was brought into the initial decision-making considerations, the National Command Authority might be more reluctant to use force.

There is credible precedence for the slippery slope argument. Many political observers believe that these operations are similar to the situation in which the United States became embroiled in Southeast Asia. First, there were advisers. Then the commitment evolved into hundreds of thousands of soldiers resulting in the more than 50,000 American names inscribed on a black wailing wall in Washington. More recently, the humanitarian operation in Somalia, Restore Hope, provides an excellent example of how even the best-intentioned efforts can change dramatically. At the onset, it was never envisioned that U.S. forces would be conducting raids to capture local warlords or be constantly confronted by snipers and mines. Rather, President Bush pronounced it "an operation to feed starving people in support of American values."

Policy decisions on the commitment of U.S. troops are always tough. Anytime soldiers are sent to a troubled area, it must be anticipated that some friendly casualties will occur. The responsibility of command is to minimize the number of casualties. Even under the most benign circumstances, accidents and criminal acts may happen, and brave soldiers may die who would be alive had they not been sent to that location. One political science professor arguing against non-lethal weapons stated that he did not want the President to have a wide range of options. "It should be kill or be killed, nothing else. I don't want him to be distracted," he said emphatically.[2]

My initial flippant response to the question of an overly ambitious President becoming careless with troop deployments was "That is job security for Congress!" What I meant was that our political system has checks and balances in authorization to send troops into combat or to other dangerous situations. It is incumbent on *both* the legislative and executive branches to discharge their respective responsibilities actively.

However, the slippery slope issue does point to a need to thoroughly examine all aspects of a situation before committing troops. It is essential that the objectives and end-states be discussed in detail, including the protracted effects on all aspects of American interests. Such comprehensive analysis and deep contemplation are generally not strengths of American politicians. They tend to be quite short-sighted, motivated by the next election, and few have the background or interest in the long-term ramifications of intricate international relations.

Use of American forces should be authorized only when we, as a nation, can articulate our interests in the situation. And, once committed, we must be prepared to accomplish the designated mission, no matter what it takes. The military manpower resources available are limited and precious. Experience has shown that doing a job halfway is worse than not doing it at all. It has also been demonstrated that the application of force has definite limitations in repression of hostilities. For example, while the former Soviet Union was able to hold the former Yugoslavia together temporarily, hostilities broke out soon after the Soviets departed, so much so that the country of Yugoslavia disintegrated. The British experience in Northern Ireland is another situation where hostilities continue even with the presence of armed soldiers.

Complex problems, such as regional stability, will take years, if not decades, to resolve. Therefore, U.S. troops must be prepared to stay longer than most Americans are prepared to commit to up-front. However, if troops are sent into sensitive situations, commanders should have the widest range of force options available. It should not be a case of "kill or do nothing." Non-lethal weapons are essential adjuncts to the field commander's arsenals.

A second concern about use of non-lethal weapons is risk of retaliation.[3] As discussed in the earlier chapters on technology, there are a number of antimatériel non-lethal weapons that could cause great damage to a country's infrastructure. Delivery of these weapons to protected targets is one of the problems faced when using them. However, the United States, and many of our First World allies, are heavily technology reliant and information dependent. Furthermore, our democratic principles dictate an openness that creates an inherent vulnerability to infrastructure-damaging weapons, especially sophisticated information warfare systems targeted against civilian computers.

The risk-of-retaliation argument urges extreme caution in developing and deploying non-lethal weapons, lest they be used against us. Some believe that our own vulnerabilities outweigh the benefits to be gained by using powerful antimatériel weapons. This argument often holds that if we do not develop these weapons, then others will either refrain from doing so, or are incapable of creating these new systems themselves.

It is true that any action taken against an adversary may be misread. In any use of force, we must be prepared to defeat an adversary who counterattacks. The risk-of-retaliation issue is not unique to non-lethal weapons. The same argument can be made for *any* action taken against a potential aggressor: show of force, or lethal or non-lethal weapons. Use of force, real or implied, must always be supported by the capability and the *will* to vanquish the enemy by whatever means are required to accomplish the task. Non-lethal weapons offer a few additional low-end options, but they are ones that may minimize long-term adverse effects from intervention.

Risk of proliferation may also be cited as a reason for inhibiting development of some non-lethal weapons. This is partially a generational issue. That is, as these new weapons become older, they may be sold or otherwise provided to less responsible organizations or governments. It is also believed by some analysts that these advanced weapons should not be manufactured, as they may be used against us. If we do not spearhead development, they hypothesize, then these weapons will not be created. However, to suggest that we are the only nation capable of developing these new weapons is another example of cerebralcentrism. There are many smart adversaries in the world. If other countries, groups of terrorists, or extragovernmental organizations perceive a need for a new weapons capability, they can and will develop it without our help. But it is true that any weapons system we design may be used against us someday. The United States has a lucrative foreign military sales program that almost guarantees that our soldiers will fight against our own systems. Therefore, to protect against this, it should be axiomatic that when development of a new weapons system is initiated, we simultaneously begin work on the countermeasures.

For many of the enemies we will oppose, matching non-lethal weapons in kind may not be an option. The question is usually asked, "What happens if we use non-lethal weapons and they shoot real bullets back at us?" This is certainly a valid concern for troops on the front line.

The intent of non-lethal weapons is to allow use of force in an attempt to prevent escalation without producing irrevocable fatalities. This situation can be handled by fully developed and fully understood rules of engagement that are backed by extensive training. As a matter of principle, non-lethal weapons should never be employed without adequate lethal support that is clearly displayed to the adversary. There must be no doubt in the mind of the aggressor that we possess sufficient force to accomplish the mission, and that we are prepared to use that force should the situation so dictate. Further, they should know that our troops are not required to use non-lethal force before shooting to kill. So the answer to the posed question is clear and simple: If they shoot real bullets, we will shoot back with overwhelming, deadly firepower.

Combat assessment with non-lethal weapons can be a serious problem. They will add another degree of complexity to an already difficult situation. During Desert Storm, penetrating, precision-guided munitions hit targets with great accuracy. The problem was that the intended damage occurred deep inside heavily fortified bunkers or several stories down from the roofs of tall buildings. The location of the damage made assessment by conventional photographic imagery quite problematic. Therefore, an urgent requirement already exists to be able to ascertain accurately the amount of damage done in areas that are obscured from external observation. One such method is to embed sensors in precision-guided munitions that independently evaluate the damage caused by the primary weapon. Evaluation could be accomplished by a staged device that would separate from the warhead prior to impact and take measurements at the time of detonation or shortly thereafter. The report would then be radioed back to waiting aircraft or via satellite to the combat assessment office.

The results of non-lethal antipersonnel weapons are fairly easy to obtain through reports from troops on the ground. However, combat assessment of antimatériel systems is a different story. For these non-lethal weapons, it will be necessary to determine not if a target has been physically destroyed but whether or not it can still function. Some of the questions that will be asked are: How much electricity is flowing? or What is the status of fuel supplies? How many trucks are capable of transporting troops or matériel? Are the communications systems destroyed or just in operational silence? How long will it be before the enemy has the capability to attack?

For most of these questions, traditional forms of combat assessment with existing sensors and platforms will be insufficient. Therefore, as new non-lethal systems are developed, there must be a concurrent effort to design sensor systems that provide the feedback required to produce combat assessments that can be relied upon. This solution will not be cheap. It will require some very advanced technology, including microrobots, nanometer-scale secure communications, unique power sources, and smart materials. However, such advanced sensors will be needed with or without non-lethal weapons.

Our ability to conduct accurate combat assessment is inextricably linked to the next critical issue—confidence. For any weapons system to be accepted by soldiers, they must have absolute confidence that it will work reliably. Part of the opposition to non-lethal weapons by troops on the ground is that they are unsure of how well the weapons systems will perform. For many decades, soldiers have used bullets effectively. As was the case when the M-16 rifle was first introduced during the Vietnam War, even new weapons that fire bullets are viewed with suspicion.[4] Still, troops understand bullets. They have trained extensively with them and they are confident of their results. Having a weapon that fires real bullets provides a sense of well-being. Now, many feel they are being asked to place their personal safety in technologies of questionable capability.

For non-lethal weapons to be accepted, it is essential that extensive training be conducted. Soldiers and their commanders must incorporate these systems into training at every level, from individual training through unit exercises, and even include them in high-level war games and simulations. Soldiers should never be handed new weapons when they are on the verge of deployment. Soldiers fight as they train. New systems of any kind only confuse an already difficult situation. However, with adequate prior training, they can be taught to use the new systems efficiently, and confidence will increase gradually as non-lethal weapons are successfully employed in real-world conflict.

Range is important to soldiers. They know how far a rifle will shoot. As seen on television newscasts, rioters can inflict substantial damage using rocks and fire bombs. New, antipersonnel, non-lethal systems must be able to reach out farther than an opponent can throw a stone. Peacekeepers must have the ability to maintain a safe distance from unruly mobs if they are to feel safe and in control of the situation. Therefore, range is important in providing confidence in the system.

At higher levels of responsibility, another form of confidence is necessary. Senior commanders need to know exactly what effect the weapons will have on targets, both personnel and matériel. To answer these questions, extensive effects testing is very important and requires much more attention in the future. However, there are many challenges to non-lethal weapons testing. Antipersonnel weapons are extremely difficult due to the sensitivity to human-subject-testing protocols.[5]

The establishment of unrealistic expectations is a major concern to both policy makers and weapons developers.[6] Talk of "bloodless war" or "war without killing" is very misleading.[7] It has already been noted that non-lethal weapons do not guarantee a total absence of fatalities. However, casual observers, often from the press, frequently do not take the time to understand the complex issues involved in the non-lethal debate. The catchy phrases tend to make the public believe one of two things: Either we will use force without risk of killing opponents or bystanders, or we have gone soft and are afraid to use our military effectively. Both thoughts are equally wrong! Anytime troops are deployed on an operational basis, it must be assumed that some deaths may occur. Further, if threatened, U.S. soldiers will respond with the force necessary to suppress the enemy.

Another category of concern is against whom non-lethal weapons might be employed. Paranoia is running rampant in the United States. We have addressed the militia movements and the surprising widespread support that conspiracy theories receive. Distrust of the government by not thousands but tens of millions of U.S. citizens is confirmed in public opinion surveys. The skepticism and controversy has been fueled by recent revelations that the U.S. government has routinely lied to the people about such varied topics as human radiation experiments, withholding treatment in the Tuskegee prison syphilis experiments, the oppressive actions of the Internal Revenue Service, the amount and geographic area covered by fallout from nuclear testing, and even UFO sightings, among others.

Many of these conspiracy theory adherents believe that the government—or some other supranational organization—is attempting to take freedom away from the citizens. Some of them see non-lethal weapons as tools to facilitate those objectives. They believe that these weapons potentially could be used to enslave them for some unstated nefarious purpose. The fallacy of this logic should be readily apparent. Sufficient force already exists to accomplish that task. Therefore, no new non-lethal weapons would be necessary. However, that does not stop the buzz on the Internet or even from being incorporated into the theme of popular movies such as the August 1997 release *Conspiracy Theory,* starring major Hollywood actors Mel Gibson and Julia Roberts.[8] There is indeed a problem. However, *the problem is a crisis of confidence,* not of non-lethal weapons technology.

There also exist a small number of those conspiracy theorists who have tied non-lethal weapons to mind control. They seem to believe that the U.S. government has an extensive arsenal of systems that can extinguish the will of any person. If the government is not involved, they then hypothesize that some mysterious supranational organization or some other order they don't know anything about is seeking to control them. Anyone familiar with the theorized capabilities would recognize these claims as preposterous. Nonetheless, these detractors vociferously express their convictions over the Internet. Conspiracy theorists notwithstanding, non-lethal weapons have nothing to do with mind control.

While the threat of domestic domination may be minimal, there is a related issue that should be acknowledged: the possibility of non-lethal weapons being used as instruments of torture. Both chemical sprays and electrical shock weapons have a history associated with misuse and torture. Unfortunately, opponents of non-lethal weapons too often greatly exaggerate emotional issues associated with the use of force designed to provide law enforcement officers with a range of options. In a 1996 television special on non-lethal weapons produced by RDF in the United Kingdom, an American woman, Loren Anderson, claimed that "In California in the past four years thirty-two people had died from pepper spray." She went on to explain that those deaths were all of people who already had been subdued and were in police custody.[9] According to Anderson's theory, these non-lethal weapons are so insidious that they will have the effect of suppressing lawful dissent by the citizenry. The reality is that there has been one death directly attributed to pepper spray by a coroner, and that was in North Carolina.[10] Further, dissent seems to be alive and well in American society. Non-lethal weapons have not changed that, nor will their future availability make any difference in the amount of dissent or how it is displayed.

In law enforcement, there is a fine line between subduing a violent criminal and inflicting unnecessary pain as punishment. An officer who has been involved in hot pursuit can easily get carried away once he or she has caught the criminal. There is no clear-cut answer, and there will always be legitimate differences of opinion concerning the appropriateness of using force. The best solution to controlling the use of force is constant training and adequate supervision.

Out-and-out torture is a different matter. Cattle prods, with their strong electrical shock, have been a favorite tool for torture for many years. Designed to herd large animals with thick skin, they pack quite a painful wallop when used on humans. Frequently applied to a victim's genitals or other sensitive locations, they can inflict almost unbearable pain that may lead to unconsciousness or, in some cases, death. Similarly, Tasers can be used to induce pain and suffering.

However, just because weapons can be misused is no reason to ban them. The North Koreans are reported to employ a simple but effective method for obtaining information from recalcitrant individuals. They use a hammer to smash a joint on the prisoner's finger. It is stated that no one has made it past three joints before cooperating fully.[11] Ice picks, vises, pliers, razor blades, cigarettes, hoses, plastic bags, pens, pruning shears, and many other implements have all been used to inflict pain on people who cannot escape. These items are but tools of sadistic persons who will always find some means to cause suffering. The argument has nothing to do with non-lethal weapons, but, rather, the intent of the perpetrator.

A similar but decidedly less valid argument concerns development of technologies that do not currently exist, but to which some moral objection is raised. Opponents of non-lethal weapons and conspiracy theorists alike lament what might be. They hypothesize about any number of nefarious weapons, including chemical and electromagnetic mind-control systems, that will create a docile society. Such systems are usually rumored to be designed and controlled by some unspecified "They" who are the ultimate powerbrokers. The "They" are often cited as the Trilateralists, Bilderburgers, Masons, Council of Foreign Relations, or other group that is not well known to those developing the conspiracy theory. Neither ignorance nor lack of information can dissuade the theorists from postulating a doom-filled scenario.

Probably the most ridiculous concern that I have heard tied loosely to non-lethal weapons is genetic engineering of soldiers. The issue was raised on national television by one of the better-known human rights advocates. He actually suggested that the government was likely to begin development of a warrior class of drones. While the theme has been used in several movies, no one else has seriously suggested such a program. Although it clearly has nothing to do with non-lethal weapons—or reality, for that matter—the argument does depict how far some people will go to evoke an emotional response in support of their cause.

There are many technologies that might be developed. The issue of propriety of any advanced technology exceeds the scope of non-lethal weapons. The head-in-the-sand approach, which assumes that if the United States doesn't develop a technology it won't be done, is specious. Banning research and development of technologies does not mean they won't exist, only that we will not have access to them. It is better to understand the technologies, at least from a defensive perspective, than to be surprised when they are used against us. Use of force is a political, not a technological, decision.

Cost of development of non-lethal weapons is a zero-sum game. In a period of limited resources, senior officials must make tough decisions about what research and development to fund and what to drop. Being relatively new in this process, the funding available for non-lethal weapons is fairly limited. One question that frequently arises is the cost comparison between bullets and non-lethal weapons. To address the issue adequately, it is necessary to evaluate more than the traditional life-cycle costs. In the equation must be the cost of limiting the response of commanders to standing aside or escalating to a shooting situation. If non-lethal weapons prevent just one situation in the next decade from becoming a hot conflict, the entire cost of R&D would be justified.

Another cost-related issue that hinders the development of non-lethal weapons is the small amount of money being put into the program by the Department of Defense and law enforcement agencies. It may be noted in the references that major defense contractors such as Lockheed-Martin, Raytheon, McDonnell Douglas, Hughes, Rockwell, ITT, and others are rarely, if ever, mentioned. Instead, it is small companies exhibiting entrepreneurship that have ventured into this market. The message from the weapons developers is clear. From a business prospective, the pot is just not big enough to make their participation a wise decision. Obviously, there must be a balance between throwing money at a problem—a solution that doesn't work—and providing sufficient funding to make the field sustainable. Given the fractionated market for law enforcement, it will be up to the DoD in conjunction with the Justice Department to convince Congress to appropriate adequate funding.

Another life-cycle-related issue is testing. There are established procedures for research, development, and acquisition of new weapons. Certainly everyone wants to ensure that non-lethal weapons are neither excessively injurious to people nor dangerous to the environment. Reasonable testing is necessary and prudent. But there needs to be a balance between operational requirements and exclusion of any and every adverse effect. After all, these are weapons systems that are being designed, not new heart valves. The standards should be prudent. Too frequently, demands are being placed on developers that are so restrictive as to dissuade them from continuing with their products. As I have stated repeatedly, when force is used, we must anticipate some level of casualties. Nothing is foolproof. A good standard that might be applied is to ask yourself, COMPARED TO WHAT? In other words, what force would be used if this new system were not available? If the new system is inherently safer, it seems reasonable to adopt it, not to demand testing ad infinitum.

In many ways, the issue I name "Compared to what?" should become paramount. Opponents of development of non-lethal weapons systems, such as the vocal critic William Arkin of Human Rights Watch, frequently distort facts, give incomplete information, or focus issues related to potential discomfort or injuries. Such is the case with his position paper denouncing all acoustic weapons. These opponents also complain about the severe stinging sensations caused by bean-bag-bullet impact, temporary dazzling, and pepper spray in the eyes of rioters. However, they totally fail to address useful alternatives. While they want—and even demand—their bodies to be property protected by law enforcement officers, they focus efforts on limiting the options necessary to accomplish that task. I submit that all of these complaints about non-lethal technologies should be compared with the damage caused by the fully authorized 9-mm round as it strikes a suspect. In most cases, that is the only option remaining. They should ask any of the "victims" they represent whether they would rather endure temporary pain or discomfort of non-lethal weapons—or be dead!

Not all encounters between law enforcement and criminals are amenable to non-lethal alternatives. There are many cases in which the offender is so dangerous or violent that lethal force provides the only viable alternative. For non-lethal weapons to be used during chance street encounters, the officers must be able to switch immediately to lethal force. No weapon currently exists that allows the rapid transition necessary to ensure the safety of the officer.

The final argument against non-lethal weapons as a military option has been addressed throughout the entire book. That position is that the United States should not engage in peace support operations. I believe the point is moot. As stated earlier in this book, we *have been* involved in peace support operations, we *are* involved in peace support operations, and we *will be* involved in peace support operations. Therefore, it is incumbent upon us to provide the full range of tools necessary for our soldiers to carry out their assigned mission. They deserve nothing less!

19

LEGAL CONSIDERATIONS

The first thing we do, let's kill all the lawyers.
—*William Shakespeare,* Henry VI

In the cold of the predawn hours the nearly exhausted soldiers huddled in the muddy trenches. Some slept fitfully, knowing that light would bring another vain attempt to dislodge the Germans who waited only a few hundred yards away. For months they had incrementally traded ground back and forth across the blood-soaked French countryside. Nearing the limits of human endurance, mitigated only by fatigue that dampens all sense of caring, each soldier experienced the knotted, astringent response in the pit of his stomach. A deathly quiet permeated this surreal realm, and the brasslike taste that comes with fear was present in many mouths. Daily, the adversaries exchanged mortar and artillery shells, followed by incessant machine-gun and rifle fire as they rose up out of the protective bosom of the earth and charged forward toward an unknown fate. Each day both sides added to the ever-mounting list of casualties.

Recently, the visceral tension had been ratcheted up with the introduction of an insidious form of warfare. In an attempt to force soldiers out of their burrows, noxious gases, such as mustard and phosgene, were being employed on the prevailing winds. Mustard gas, a particularly nasty agent, blistered victims horribly by chemically extracting the body's own fluids, sometimes literally drowning the victim. Other chemicals that destroyed the blood supply and nervous system were also being introduced. These dreaded, silent killers would drift into the trenches, there to choke the unprepared. Those exposed died an agonizing death. Survivors did not fare much better. While they might recover from the painful initial wounds to their bodies, chances were that, due to damage to lung tissue, they would never be able to breathe properly for the rest of their lives. Others would be blinded and disfigured.

Thousands upon thousands were stricken with these deadly agents during World War I. The soldiers and their families would bear the burden of exposure to chemical weapons for the remainder of their lives. Faced with graphic images of the horror brought on by the gases, most countries signed the 1925 Geneva treaty to ban their use in future wars.[1]

The notion that chemical warfare was inherently evil and explicitly prohibited by treaties was one of the first issues raised when I began promoting non-lethal weapons.[2] The lingering memory of the casualties of poisonous gases used in World War I has been indelibly imprinted on the psyches of the military, policy makers, and antiwar advocates. While all agree that antipersonnel chemical weapons that produce exceptionally painful injury with long-term debilitation have no place in modern warfare, the sweeping condemnation of all chemical weapons belies two important issues. First, not all chemicals produce injury. And, second, the question "Compared to what?" must be thoroughly considered. Both will be addressed in this chapter.

The attempt to regulate warfare and introduce a degree of humanity is not new. However, when we review the results of those efforts, it is hard not to come to the conclusion that *law of war* is an oxymoron. These laws are written by lawyers and politicians who rarely get their boots muddy, come face-to-face with an adversary bent on ending their life by any means possible, or experience the ineradicable memory of reassembling explosively dismembered buddies and placing them in body bags. They have created a strange juxtaposition of altruism, pragmatism, and sophistry that has produced the complex, almost unintelligible rules by which conflict is to be played out. Unfortunately, these treaties and laws do not apply equally to all with whom we might engage. The rules of engagement for flights over Bosnia became so complicated that jokes were made in NATO suggesting adding a third seat in each aircraft so that a lawyer could fly as an additional crew member.

The first weapon to be formally banned from the battlefield was the crossbow. In 1139 the Catholic Church's Second Lateran Council decided that the crossbow caused excessive injury and could not be used between Christians.[3] The ban was not total, as it permitted the use of crossbows against those classified as *infidels.* The prohibition fell apart when Richard I of England did not discriminate between infidels and the French.

Later, exploding bullets became a concern. Designed to destroy hard targets, the bullets would cause massive internal damage to humans hit with them. Used by the American Army in the Civil War and the British Army in India, other nations worried about more extensive applications on battlefields. In 1868 the Russian government convened an International Military Council, which drew up a document that is known as the *Declaration of St. Petersburg.*[4] The prohibition against exploding bullets—those less than 400 grams—and other special rounds such as what are known colloquially as dumdums, still continues. In fact, when the Colt AR-15 rifle, forerunner of the now standard infantry M-16, was first tested, the rounds were found to be only marginally stable in flight. One concern was that the tumbling rounds would cause excessive physical damage and the weapon might be declared illegal. Increased stability was achieved and the weapon fielded.

Over time, there have been a number of declarations and conventions that have some applicability or relationship to non-lethal weapons. A list compiled by Lewer and Schofield of Bradford University in the United Kingdom includes the following:[5]

- The Lieber Code—1883
- Declaration of St. Petersburg—1868
- Hague Declaration (IV, 2) Asphyxiating Gases—1899
- Hague Declaration (IV, 3) Expanding Bullets—1899
- Hague Declaration (IV) Laws and Customs of War on Land—1907
- The Protocol for the Prohibition of the Use in War of Asphyxiating Poisonous or Other Gases, and of Biological Methods of Warfare—1925
- Geneva Convention—1949
- Convention on the Prohibition of the Development, Production, and Stockpiling of Bacteriological (Biological) and Toxin Weapons and on their Destruction—1972
- Convention on the Prohibition of Military or Other Hostile Use of Environmental Modification Techniques—1977
- Geneva Conventions, Additional Protocols I & II—1977
- United Nations Inhumane Weapons Convention—1980
- Chemical Weapons Convention—1993

While aspects of these declarations and conventions have proven beneficial in some measure during the conduct of war, none has been foolproof. Overall, the plight of prisoners has been improved in many conflicts. However, the brutal interrogations and wanton torture of U.S. aviators shot down over North Vietnam serves as a clear indicator that even under external scrutiny, illegal practices continue. Chemical weapons are still being stockpiled, to be employed by rogue states and terrorist groups. Even signatories to conventions have broken their word. Immediately following the 1972 Bioweapons Treaty, Leonid Brezhnev instituted the largest biological weapons program the world has ever known at Novosibirsk, Russia. Despite the treaty, the illegal work was established at a minimum of forty-seven sites. Among their accomplishments was a genetically modified strain of anthrax that was resistant to all known vaccines and antibiotics.[6]

Any lingering doubt about the abject disregard for the Bacteriological (Biological) and Toxin Weapons Treaty was put to rest with the defection of Dr. Kanatjan Alibekov twenty years later. Now known as Ken Alibek, the second in command of a branch of the Russian BW program, he has described in frightening detail how each major U.S. city had been targeted with ICBMs with warheads containing a "cocktail" of agents. These mixtures included combinations of disease such as plague, anthrax, and previously unknown forms of smallpox. In fact, he even described how smallpox and Ebola, as well as other deadly agents, might be crosslinked. Through a simple process of diversifying agents, they made immunization against all physically impossible. Dr. Alibekov noted that they had made a sufficient amount of BW agents available to "kill the entire population of the world many times over."

Of course, Russia now denies these allegations. Oleg Ignatyev, a top biological expert, claims that the programs were terminated in 1990 on the orders of Mikhail Gorbachev. Alibekov states that he was working on the program in 1991, and that vestiges probably continue today. At best, we are left with believing a representative who defied the treaty for eighteen years. At worst, we are woefully unprepared for an attack that most certainly will occur. Either way, this gross violation vividly demonstrates the level of reliance that should be placed on unverifiable and unenforceable treaties. While such treaties may make their authors feel good about their altruistic contributions to humanity, the downside is potentially disastrous. The most effective deterrent to their use has turned out to be fear of retaliation, not concern for contravention of some treaty or international law.

Military casualties are one thing, but the rise in collateral civilian fatalities is of particular concern. In our first experience with war, the American Revolution, civilian casualties were almost negligible. However, the number has constantly increased. In World War II there was a dramatic increase in attacks on areas occupied primarily by civilians. Some cities were bombed specifically to demoralize the population. When reviewing all wars in the world, by the 1950s, noncombatants accounted for approximately half of all casualties. By the 1980s that number had increased to nearly 80 percent.[7] The U.S. experience of the 1990s brought about a dichotomy. The precision weapons used in Desert Storm kept collateral casualties to an absolute minimum. Conversely, in the humanitarian operations in Somalia, because of the sniper problem, the number of civilians killed skyrocketed, and was probably in excess of 80 percent. If the genocide perpetrated in Bosnia and between the Hutus and the Tutsis in this decade is factored in, the percentage of civilians killed would probably exceed 90 percent of all casualties. Clearly, there is a need to do everything possible to reduce the needless extermination of noncombatants. Holding those responsible legally accountable for their actions is only part of the answer. The number of noncombatant casualties suggests a necessity for introduction of more non-lethal alternatives.

Since the United States chooses to accept the moral obligation imposed by the declarations and treaties, it must abide by the tenets set forth. As a matter of Department of Defense policy, a legal review must occur before any contract is awarded for engineering or manufacturing development of a new weapons system. As the U.S. Marine Corps has been appointed the executive agent for non-lethal weapons, it is the responsibility of the Navy Judge Advocate General—better known by the acronym JAG—to conduct the legal reviews.

In so doing, there are three major issues pertaining to the law of armed conflict to be examined for each and every new weapon proposed:[8]

1. Does the weapon cause suffering that is needless, superfluous, or disproportionate to the military advantage reasonably expected from the use of the weapon?
2. Can the weapon be controlled so as to be directed against a lawful target and be discriminate in its effects? and
3. Are there any extant rules of law that prohibit its use in the law of armed conflict?

The intent is to ensure that there is a balance between suffering and military necessity. Normally, the principles involved include prohibitions against unnecessary suffering, indiscriminate effects, and treachery or perfidy.[9] In general, most non-lethal weapons meet all of these tests. Certainly, when one considers the alternatives, it seems preferable to incapacitate temporarily than to kill. However, there are many who believe that the old-fashioned way—to puncture, dismember, or fry an opponent—is the more desirable option. At least it is better understood. And it is this level of understanding and established legal precedence that provides a degree of comfort not available with many non-lethal weapons. As conditions have changed, so must our legal imperatives.

With the fielding of non-lethal weapons, some of the arguments already have been addressed. Three technical areas generate the most emotional debate: chemistry, biology, and lasers. My first face-to-face encounter came at the first meeting of the Council on Foreign Relations Study Group on Non-Lethal Weapons. After I gave the opening presentation, an overview of non-lethal weapons, Myron Wolbarsht of Duke University, a vocal opponent of lasers on the battlefield, talked about the dangers of antipersonnel laser weapons. He invoked an emotional appeal against blinding lasers and suggested the U.S. military was developing such systems. In my final assignment as Director of Advanced Systems Concepts, in the Army at Laboratory Command I had been responsible for the research and development of all tactical directed-energy weapons. None of the weapons described by Wolbarsht was ever considered for development. It was well understood that lasers that intentionally blinded soldiers would be illegal. When I advised Wolbarsht that he was incorrect about his assumptions that the military was developing such weapons, his response was amazing. He responded, "But you might."[10] There are many things that are technically feasible that we don't do for legal and policy reasons. The policy position prohibiting the development of lasers that intentionally blind soldiers continues to be in force.[11] However, that does not stop opponents of the field from continuously raising the same issues that have previously been definitively answered.

There have been several attempts to control lasers on the battlefield. The Red Cross has taken up the issue and produced publications about blinding laser weapons.[12] From a military perspective, laser weapons have a number of benefits. The laser travels at the speed of light and, based on the size of the power source, laser weapons can fire many times faster than bullets. Even the Red Cross acknowledges that antisensor lasers on other antimatériel applications of lasers are considered legal.[13] However, some opponents want all lasers that have the *capacity to blind* to be eliminated from the battlefield.[14] Unfortunately, that would include the range-finders on the M1 Abrams tank and guidance for some precision weapons. That is not going to happen. In response to the mounting external pressure, the Office of the Assistant Secretary of Defense (Public Affairs) issued a press release noting that "laser systems are absolutely vital to our modern military." It went on to say, "They provide a critical technological edge to US forces and allow our forces to fight, win, and survive on an increasingly hostile battlefield."[15] This comment preceded the adoption of new protocols regarding lasers by the International Committee of the Red Cross on 13 October 1995. The protocols were limited to lasers that intentionally blind soldiers not using optics and were supported by the United States.[16] It should be noted that use of a laser to blind does not constitute unnecessary suffering, nor is it illegal. This legal opinion was authored by the Judge Advocate General of the Army and concurred with by the Judge Advocates General of the Navy and Air Force.[17] They specifically noted that "battlefield commanders and laser device users have the right to assume the lawfulness of the laser devices."

The bottom line on lasers for military applications is that they are legal on the battlefield provided their primary purpose is not to *intentionally cause permanent blindness* of enemy soldiers with *unehanced vision.* Range-finders and antisensor laser weapons meet the test of legality. It is acknowledged that eyeballs behind sensors that intensify light are at risk. Most non-lethal optical munitions will probably be permissible.

In law enforcement, a very different laser has emerged. Low-energy lasers can be fitted to pistols and rifles to enhance accuracy. Usually, they display a visible red dot designating the point of impact. Some police departments have prohibited the use of these aiming devices. The fear is that lawsuits might be filed if officers fatally shoot a criminal instead of attempting to disable him or her. It could be argued, the rationale goes, that death was an excessive use of force because wounding would be possible.[18] Some police agencies do allow the aiming devices, as they reduce the possibility of hitting the wrong person.

If you think legal issues concerning laser weapons are confusing, wait until you see what chemistry has to offer. One of the first letters to the editor regarding non-lethal weapons bemoaned their use, as they were chemical in nature, and therefore must be illegal. The response from a Washington lawyer was quite amusing. He noted that because of poisonous chemicals contained in cigarettes, he would be more concerned if we dropped them on the enemy than if we used non-lethal weapons.

The use of some chemical weapons depends upon the circumstances involved. The best known chemical agents are designated *riot control agents* (RCAs), sometimes referred to as tear gas by the public and news media. Riot control agents are currently prohibited as a *method of warfare.* However, they are legal for use by law enforcement agencies, including the purpose of controlling domestic riot. Under the Chemical Warfare Convention (CWC), signed 13 January 1993 and ratified in 1997, RCAs can probably be used during peacekeeping operations, but not if an engagement escalates to international conflict.[19] Of course, the CWC is written in arcane language so complex as to bewilder the average Mensan. The reality is that the CWC has provided job security for scores of lawyers.

In her excellent paper on non-lethal weapons, Lt. Col. Margaret-Anne Coppernoll reports a conversation with a Navy JAG officer regarding the rationale for banning RCAs from the battlefield. The JAG officer stated that such an agent was "unacceptable in armed conflict because it could easily be confused with chemical weapons of a more lethal nature by the enemy who could then be provoked into escalating the conflict via a retaliatory response."[20] Of course, the confusion argument is specious but indicative of the problems encountered by warfighters attempting to do their job while encumbered by pococurante bureaucratic overseers. The same argument made by the JAG could be used for smoke and related obscurants, which are routinely employed in warfare to permit armored forces to maneuver and destroy an enemy. In fact, soldiers are taught to put on protective masks whenever smoke is encountered specifically because chemical or biological agents could be embedded. The notion also fails to take into account that agents can be designed that are colorless, odorless, tasteless, and extremely deadly. Escalation is a risk of any action taken. Military commanders should always anticipate the possibility that an adversary may overreact or retaliate and have sufficient force available to overcome that threat. That is the nature of conflict, and is no reason for prior restraint.

Barrier foams are equally ambiguous. The foam reviewed by the Navy JAG contained a riot control agent additive. Therefore, it could not be used against combatants, or in situations in which combatants and noncombatants were commingled. However, it probably could be used in a rescue operation where there was no combatant adversary. Additionally, foams can be tailored and do not require RCAs to be effective.

On the positive side, it does appear that most antimatériel non-lethal chemical systems would meet the tests of legality. In general, they are designed to attack a small portion of the target, therefore they are discriminate and frequently nontoxic.

However, they must also be environmentally safe to comply with yet another treaty, the 1977 Environmental Modification Convention (ENMOD).[21] In this document, environmental modification is defined as "changing through deliberate manipulation of natural process the dynamics, composition, or structure of the earth, including its biota, lithosphere, hydrosphere, and atmosphere, or of outer space." The convention prohibits widespread techniques, which are defined as covering several hundred square kilometers; being long lasting, meaning several months; or being severe. One wonders if this convention could be used as a rationale to stop large-scale deforestation.

Even in my earliest presentations, I noted that the United States was unlikely to develop weapons that were harmful to large segments of the environment. I would say, "No one wants to be responsible for the next Agent Orange." One day, when I was speaking before a group from the National Academy of Sciences, a hand was raised. When I addressed him, the former senior government official said, "Would you like to meet the man who approved Agent Orange?" He went on to discuss the testing that had been conducted prior to fielding the famous herbicide used to clear the jungle back from the roads in Vietnam. Certainly they had considered the long-term effects. Their best judgment was wrong, demonstrating the difficulty in predicting all possible ramifications of new systems.

The third controversial area is biology. By fiat, anything connecting biology and war must be bad. The 1977 Convention on Biological and Toxic Weapons condemns all biological agent or toxins "of types and in quantities that have no justification for prophylactic protection and other peaceful purposes."[22] The problem that has been previously noted is the almost exponential increase in the development and use of biochemicals in remediation processes. The use of these agents varies only by intent, which, as we have seen, is a critical legal issue. It means that applications will have be evaluated on a case-by-case basis. Given the emotional impact of the words *biological warfare,* you can bet that we will prohibit development of such systems. Others, I believe, will not be so constrained.

Other non-lethal weapons systems that must be considered include low-impact kinetic rounds, acoustic systems, and electromagnetic weapons other than light. Before deployment to Somalia, a number of non-lethal weapons were reviewed and approved for use including several low-impact rounds, sting grenades, and sticky foam. Each passed the test. Acoustic weapons are questioned because of the area effect, but advances in directionality should meet the requirements for fielding. Harmful interference with radio services or communications is prohibited under the Nairobi Convention. It does not apply during war, and electronic warfare is a fairly mature art. Most of those legal issues have been worked out.

Use of non-lethal weapons by law enforcement agencies brings another set of legal concerns. Liability is the major issue whenever force is used. Given the rate of suits brought against municipal, state, and federal government agencies, they have reason for concern. There is a paradox with these new weapons. On one hand, bullets are well understood, and legal parameters for their use are firmly established. On the other hand, having non-lethal weapons, even though not as well known, may result in suits either for excessive force if they are used, or for shooting if they are not used.

The use of computer viruses and other information warfare techniques is another burgeoning field requiring legal review. The law in this area is emerging at this time. It will be years before the issues are satisfactorily resolved. Most believe that hacking is a crime and there are many laws to cover breaking into another's computer.[23] International laws and rules of engagement regarding IW in actual conflict are less clear than in domestic situations. Difficult as it may be, the technicians involved in information warfare are attempting to discern the legal standing. For instance, can a stealth virus, legal during periods of armed conflict, be implanted before the onset of hostilities? What discriminates between civilian and combatant targets when facilities are shared? What constitutes legal surreptitious entry into a potential adversary's computer network versus hacking? How does that apply in domestic versus international networks in which part reside in the United States?

In addition to legal issues about technology, there are equally thorny administrative legal problems. One that has been raised contends that if lethal weapons are employed when non-lethal ones are available, it might constitute excessive force. The United States has gone on record stating that because we have non-lethal weapons, that does not infer that we must use them prior to escalation to lethal weapons. We reserve the right to use the force determined to be necessary to accomplish the mission. No warnings are necessary.[24]

This basic precept of ability to discriminate necessitating use of advanced weapons systems has already been tested and found to be without merit. Ridiculously, after Desert Storm, a peace group attempted to argue that "dumb bombs" were illegal, as they are indiscriminate. Since precision-guided weapons are available, the rationale went, we were obligated to use them and not drop our conventional munitions. Just as a myth exists that *police must fire a warning shot,* there is no moral or legal necessity to escalate force options gradually by using non-lethal weapons before lethal weapons are employed. Common sense will never stop lawyers in pursuit of a cause.

The fundamental legal problem that exists in use-of-force issues is that such problems are defined in terms of technology, when the real culprit is people and their intent. It is easy to develop an emotional argument against a technology or a class of weapons. The most notable recent example is the move to ban land mines. There are an estimated 26,000 casualties from mines each year. Based on horrific pictures of men, women, and children suffering from traumatic amputations of their limbs, it is easy to generate support for a ban on the weapons that caused them. Deeply touched by the plight of these unfortunate victims, Diana, Princess of Wales, was a high-profile spokesperson for the cause. Then, with her untimely death, the land mind issue was catapulted even farther into public view. So much so that the Nobel Prize for Peace in 1997 was awarded to the groups that led the effort to enact a treaty to ban such weapons. While organized with the best intentions, the movement focused on the wrong problem. The real problem was indiscriminate use of explosive devices that resulted in those maiming injuries and deaths. In many cases, it was mines that facilitated the tragedy. However, someone put them there.

In Vietnam, mines and booby traps caused a large number of casualties, and I had firsthand experience with them. On at least one occasion I came within millimeters of being one of those casualties when my foot struck a tripwire and began to pull the triggering device. Just in the nick of time, a

Vietnamese lieutenant saw the camouflaged booby trap and stopped me from blowing my legs off. However, what we feared most was not technically a mine, but rather unexploded cluster bombs. Known as CBUs, these antipersonnel, high-shrapnel-content bomblets were dropped from American planes and sometimes failed to detonate on impact. Collected and reconfigured by the Vietcong, they made nasty booby traps that perforated the legs and lower bodies of many soldiers and civilians. The point is that, by definition, the treaty would not ban the use of CBUs because they are not designed as mines but rather as bombs. However, the treaty would ban the use of mines that could be safely employed through the introduction of timing mechanisms that render that explosive ineffective after a designated, and relatively short period of time. While this technical solution was proposed by the United States, it was rejected and the broad language imposed.

Opposition to various technologies is intentionally stated in emotional terms. When logical arguments are made against the prevailing opinion, the response is usually an ad hominem attack. A classic example is the comments made by Jody Williams immediately after it was announced that she was the figurehead for the receipt of the Nobel Prize. Since the United States had not agreed to the treaty in its current form, she stated, "President Clinton would find it hard to say he was a leader." This was based purely on the notion that President Clinton took a position different from hers. Williams continued by stating that a vote against her position was "against humanity" and that the President "did not have the courage to be the commander in chief of the military."[25]

The reality is that there are military operations in which mines are necessary. As a former special operator, I am very sensitive to their utility in breaking contact when a small unit is being closely pursued. They can also be employed to create barriers and protect exposed flanks of one's position. Properly emplaced, minefields are clearly marked and mapped. It is the improper use of mines that has caused the majority of the collateral casualties. Further, in situations such as breaking contact by small units, the timing mechanisms proposed can adequately solve the problem by rendering the mine inert in a period of hours.

Actually, the land mine issue has a direct bearing on non-lethal weapons. Years before the treaty was presented, the United States had taken a position that would attempt to find non-lethal alternatives to explosive antipersonnel mines. However, with broad-sweeping language, even advances designed to meet the ultimate aim of the negotiations—the reduction or elimination of traumatic amputations and fatalities from mines—could be deemed illegal. Again, there is a need to identify and solve real problems. Blaming technology for social issues may be easier than addressing the root cause. However, such legislation will not yield true resolution. In fact, it may serve to mask the issue.

In a single chapter it is impossible to review all of the legal ramifications of non-lethal weapons. Key issues often are blurred. It is not clear exactly what constitutes armed conflict, or how to discriminate between civilians and combatants. Since intention is in the mind of individuals, it can only be inferred based on actions that are observed. Similarly, how to judge proportionality is open to interpretation. Treaties and legal policy too frequently attempt to solve the wrong problem. By isolating technologies and defining them as *the problem,* we miss the point that people are really at fault. We define the issue in terms of technology because it seems easier than dealing with the root cause. Overly simplified, this is analogous to "Guns don't kill people; people kill people." With all non-lethal weapons technologies it is the misapplication that is of concern, not the basic technology itself. Most of the technologies used in non-lethal weapons have other, peaceful purposes. The genies are out of their bottles and they won't go back.

There are no simple answers. Current conventions, however altruistic and well intentioned, are fundamentally flawed as they are based on the concept of war between nation-states. The reality is that such agreements may constitute multilateral disarmament while disregarding several current and emerging nonstate adversaries, each possessing the capability to inflict great injury on the United States and its allies. Some of those future antagonists would have no compunction about blatantly violating covenants to which they are not a party. An asymmetrical power differential exists. It should not be needlessly relinquished.

Given the change in the nature of conflict in many areas, limited or indistinct objectives, and rapid advances in technology, treaties and conventions that have already been agreed to may need to be reexamined. That position will not be popular with organizations that have established careers based on developing the agreements and enforcing them. A critical issue in future weapons system development should be a comparison of alternatives as they affect desired outcomes. If the laws turn out to be wrong, change them. We should not be tied to emotionally based, anachronistic circumstances. In today's vernacular, that translates to "Do the right thing!" The United States can, and should, maintain a position on the moral high ground. In so doing, it must be acknowledged that our actions will impose certain risks on our troops and our citizens. We should minimize those risks, but our value system dictates that we follow a path that will periodically employ force when we believe it to be right. However, we should also be both pragmatic and cautious. Teddy Roosevelt's admonition to carry a big stick is still good advice.

When determining when it is appropriate to use force, future decisions will probably be more difficult than past ones. It will be necessary to look at many of our objectives quite differently. What will come under intense scrutiny, and the issue we will explore next, is the fundamental concept known as winning.

20

WINNING

Failure is not an option.
—*Ed Harris as Flight Director Gene Kranz in* Apollo 13

Winning is the American way. Next to freedom, it is the most cherished value in the country. Winning is good. Winning big is better. "Winning isn't the most important thing; it is the *only* thing." That famous quote attributed to Green Bay Packers football coaching legend Vince Lombardi resounds throughout every locker room in America. Each year in the National Football League, only two teams make it to the Super Bowl. One becomes known as the winner, and the other is quickly forgotten. The same is true for professional baseball when two teams play in the World Series, and the National Basketball Association and the NBA Playoffs. Second place is just not good enough. Only the winner is recognized and remembered.

It is not just in sports, but in every aspect of our lives, that we embrace the concept of winning and losing. We compete for grades in education, for money in our occupations and economic endeavors, and for power in politics and personal relationships. Our legal system is, by design, adversarial. Regarding your health, we battle to conquer disease while some even vainly attempt to evade death. Beyond just striving to do better, our gain is usually at the expense of another person's loss.

National security traditionally has been defined by winning and losing. We won World Wars I and II. We won the Persian Gulf War. We talk of winning the Korean War, even though it ended in a stalemate at the 38th Parallel where North Koreans face toward the south with guns pointed at us. We try not to talk about the outcome in Vietnam. However, the most important conflict in modern history did not take place on the battlefield. It was the Cold War, punctuated with perilous episodes, such as the Cuban Missile Crisis, that on more than one occasion brought us to the brink of global destruction.

The similarity in all of these hostilities was that there was a visible and easily defined enemy. For all of the flash points around the world throughout the Cold War period, one compelling fact remained: Because of the Soviet nuclear arsenal with their long-range delivery capabilities, our *national survival* was at risk. In that environment of oppressive intimidation, truly, failure was not an option.

Then, with the precipitous political demise of the Soviet Union, that threat to our national survival was dramatically diminished. Almost instantly, the bipolar global identity that had defined the world for more than four decades ceased to exist. One superpower remained. New alliances were formed, nations dissolved, and others evolved. Power vacuums were tested and filled. Geopolitically speaking, many areas of the world were unstable or near chaos.

In reality, the precataclysmic world had provided our defense planners with an ability to focus attention on a strong, clearly defined adversary. The existence of a predominant enemy was very advantageous, as military contingency objectives could be generated with precision. Sweeping directives were promulgated: Ensure the survival of the United States. Prepare to restore the geographic boundaries of Western Europe. Ensure that the sea lanes are kept open. These are examples of the strategic-level military objectives that were used as a means to develop and resource our military forces. While other military operations were contemplated, they were basically viewed as "lesser included" actions. It was believed that if the military was designed and structured to meet the most dangerous threat, then it should be able to accomplish any mission requiring a limited projection or application of power.

During the Cold War years, resourcing the military was the fundamental issue. Given the hypothesized size and capabilities of the combined Soviet and Warsaw Pact forces, very large, highly sophisticated military units equipped with the most advanced technology were required to counter them. Planners always thought we needed more: more men, more tanks, more aircraft, more ships.

When concerned with national survival, Congress, with the support of the American voters, was prepared to appropriate very large budgets to meet the military's requirements. While the size of the defense budget varied, it always absorbed a significant portion of the American federal budget. During the early years of the Reagan era, the dollars allocated to the Department of Defense soared. Some analysts suggest that it was that buildup in the early 1980s that brought the Soviets to their knees. It is conjectured that we just outspent them. Since the collapse of the Soviet Union, the Defense budget has steadily declined until we have reached a point at which there is concern about the adequacy of the force to protect our national interests. It is the combination of an astatic world structure and our diminishing military resources that requires us to reevaluate our objectives in application of force. Those issues will even cause us to redefine our concept of winning.

While winning is important, how victory is accomplished is also of consequence. In American consciousness we differentiate between winning big and a massacre. This is inherent in our concept of fairness. As previously stated, winning by a substantial margin is considered to be good. However, participating in a perceived massacre is unacceptable.

The gap between the two is narrow and blurred. Revisionist history frequently shifts battles from one category to the other, depending on the current view of propriety. As an example, atomic bombs were dropped on both Hiroshima and Nagasaki, Japan, leading to an abrupt end to World War II. At that time, almost no one questioned the moral decision to use the world's first nuclear weapons. The strategic objective, to obtain unconditional surrender at the earliest possible time, was accomplished. We won and won decisively. However, fifty years later, a great debate has arisen about whether or not dropping the bombs on populated areas was the right decision. Employing 20/20 hindsight, altering the facts, and disregarding the mood of the world, revisionists raise superfluous questions that cannot alter the course of history but may impact future decision making. The people who should answer the question of the appropriateness of dropping those bombs are the soldiers, sailors, marines, and airmen who were gathering to form the largest invasion force ever contemplated. They are the ones who would have been among the estimated 1,000 casualties per day for the ground invasions of the Japanese homeland.

Another, more recent, example of the difficulties that can be encountered in making moral decisions occurred in February 1991. It came on the fourth day of the ground battle during Desert Storm, as the defeated and demoralized Iraqi soldiers attempted to flee Kuwait City. After looting the metropolitan area of easily transportable valuables, they took their armored vehicles, along with any other conveyance they could lay their hands on, and struck out for Iraq. However, the UN air forces were waiting in ambush for them. First, the fighters destroyed the lead and trailing vehicles. Further constrained by the vast desert surrounding them, this action effectively blocked vehicle movement in any direction. What followed was called a *turkey shoot* by the television reporters who were able to obtain pictures of the carnage that unfolded. Wave after wave of fighters and gunships systematically destroyed the trapped vehicles. So much destruction was inflicted that the incident became known as the *Highway of Death.* Intense firepower was directed against the hundreds of vehicles that lined the miles of thin road that stretched toward Baghdad. Iraqi soldiers fleeing the scene on foot mercifully were allowed to leave without being gunned down. However, they had to cross the expansive barren wasteland to reach safety.

Chairman of the Joint Chiefs of Staff General Colin Powell, Secretary of Defense Dick Cheney, and President George Bush all knew that the televised images of death and destruction were too much. They could not allow the world to believe that our troops would wantonly kill nearly defenseless enemy soldiers who were only attempting to flee with their lives. The strategic decision had already been made that no ground assault would be made to capture Baghdad. Television broadcasts around the world had picked up the story of the Highway of Death. To avoid continuance of action that would soon be perceived as a massacre, President Bush announced that the campaign objectives had been met, thereby effectively ending the war.

This decision was a difficult one. It was openly questioned by military analysts, as it left the Iraqi military with a fairly large armored force intact. The intent was to prevent a military vacuum in the region. Political leaders did not want so many of Saddam's tanks destroyed that Iran might be encouraged to invade Iraq again. But they wanted to leave him only sufficient combat power to defend his country, not enough for him to be an offensive threat in the region. Unfortunately, a substantial part of his better-trained and -equipped Republican Guard withdrew before it could be destroyed in the remaining time allotted before the cease-fire went into effect.

Our national survival remains the number-one military objective and seems assured for the immediate future. The core objectives of current U.S. security strategy include effective diplomacy supported by military forces prepared to fight and win, bolstering America's economic prosperity, and promotion of democratic principles.[1]

However, the functional strategy has become one of regional stability, and other, more abstract objectives are articulated, including preventing transfer of military critical technologies relating to weapons of mass destruction, countering terrorism, reducing the flow of illegal drugs, and preventing damage to the environment.

As a result of these new objectives, our strategic considerations for projection of power and application of force have changed. It is necessary to determine how we will protect our national interests and values. Therefore, the requirements placed on our military units must also change. In addition to maintaining our ability to project and apply power to achieve domination of an adversary, the United States has found itself engaged in many small military operations—ones where using physical power is not the main issue—nominally, operations other than war. Such examples as Haiti, Somalia, Panama, and Bosnia have already been discussed.

Not all in the military agree that they should participate in such operations. Many young commanders find this concept for use of force an anathema. They prefer to be given clearly defined, obtainable military objectives, then allowed freedom to accomplish their mission. In Army jargon the operations order should be "Three up, two back, hot socks on the objective."[2] This is the way soldiers have been trained to think. Give them the order to take the high ground, secure the sea lanes, or dominate airspace, and they will do it. Politics should be left to politicians. As far as many junior leaders are concerned, once military forces are committed, they should be free to destroy the enemy. This is now wishful thinking on their part. The world has changed, and so too must our concept of use of force.

Under conditions that approximate all-out traditional warfare, the old definition of winning will continue to be respected. Such circumstances, in which large land, air, and sea forces are decisively engaged in deadly combat with a determined enemy, have not been eliminated from the range of possible missions. While the probability of such engagements has been reduced, preparations for such an eventuality are appropriate.

However, there also must be definitions of winning that do not include holding the severed head of your adversary for all to see, or raising your flag over the enemy's capitol. Recent engagements have presented a different reality of conflict, one that requires a new definition of winning. In the future, "winning" may be more amorphous than in the past. Winning may have to be defined in such terms as reestablishment of regional stability. It might also have a temporal component and be defined in terms of periods of relative quiescence or a lack of hostile actions.

Winning may be accomplished when an opponent meets specified demands, such as keeping its troops within designated boundaries, providing humane treatment to minority groups, or allowing economic trade to occur. The movement of refugees and migration of disenfranchised people will become an integral aspect of future stability. All of these objectives may be accomplished without physically conquering the enemy and occupying his territory. Similarly, in the future, peace also will be seen in provisional terms. These terms may include a relative absence of armed conflict or terrorism. Peace may also encompass freedom of movement, acknowledgment and protection of human rights, and the unrestrained ability to express and communicate thoughts and ideas. Future considerations may even include maintenance of an environmental balance or some minimum standard of living for all members of our species.

In responding to infractions, the concept of proportionality has become an issue that requires far more attention at all levels of command. Proportionality is a concept in which the force used is limited to an amount that is only necessary and proper to counter an aggressor's actions. Those very senior Defense officials who engaged in planning for the use of nuclear weapons did consider this proportionality as a threshold issue. Except for a few isolated incidents, it was not a concern of most field commanders, but articulation of proportionality was largely a point of debate for lawyers and policy makers. While it has long been recognized that you don't use a sledgehammer to kill flies, proportionality as a key issue in force generation has only recently been discussed specifically by military commanders at all levels. Now, however, proportionality is important when contemplating the measurement of achievement objectives.

Non-lethal weapons could play an important role in providing response options when a minimal infraction has occurred, one that cannot be ignored for fear of further escalation. They would provide senior decision makers with the opportunity to damage property, thus sending the message that violence will not be tolerated, without crossing the proverbial *death barrier.*

Redefinition of winning is paramount when the vital issue of resourcing our military forces is addressed. Maintenance of troops deployed across the world is extremely taxing. While the number of soldiers committed at any location is substantially smaller than those provided in a major campaign, the cumulative effect is substantial and draining. Further, achieving objectives such as ensuring regional stability requires extended obligations. The United States is not known for support of long-term commitments. In the future, therefore, our politicians must thoroughly contemplate the consequences of the application of force.

Another important consideration is that there are decreasing numbers of soldiers from which to draw. Some units are already finding themselves being rapidly redeployed from one location to another. The impact on morale and family life should be obvious. Training and readiness for major combat operations also suffer in this environment. A recent Army survey reported that half of the respondents thought the service had become *hollow* again, and was overtaxed by foreign deployments.[3]

Given the complexity of future interventions, it is imperative that we provide our troops with the weapons necessary to carry out these complex and often sensitive missions. Many scenarios demand that application of force be limited. If the mission is to establish or reestablish stability to a region, contributing to the body count is rarely efficacious. Therefore, conventional weapons, while essential for troop safety, are of limited utility in accomplishing missions. Conversely, while counterintuitive to many, non-lethal weapons are the tools necessary to achieve the stated objectives.

Missions on the magnitude of a major regional conflict can also benefit from the availability of non-lethal weapons systems. In examining the concept of strategic paralysis, it was demonstrated that these weapons could be instrumental in degrading or destroying infrastructure targets. When attempting to minimize collateral casualties in areas heavily inhabited with noncombatants, they provide a unique advantage that cannot be gained even with precision-guided weapons. Such controlled strikes can be instrumental in procuring support of that population and reestablishing peace or stability.

Even during mid- to high-intensity battles, non-lethal weapons may play an important role. They can be used to gain and hold terrain that is difficult for troops to occupy. By cutting down on casualties whenever possible, these weapons can assist in the enemy's acceptance of terms for termination of conflict while minimizing resistance and animosity that destabilizes the post-conflict situation.

Acceptance of alternative definitions to winning will be extremely controversial and fought long and hard. Traditional enemies will remain, and we know how to deal with them. However, in a world where some of the new adversaries don't have addresses, some situations proscribe the use of lethal force, and some of the offenders are us, the infantry concept of "closing with and destroying the enemy" has little meaning or validity.[4] For those circumstances, conventional force cannot be used effectively. However, it is just such situations, undesirable as they may be, that will continue to generate new requirements for troop deployment. They also indicate the necessity for more non-lethal weapons.

The reality is that the nation, and the military, are resource constrained. The same is true for our allies. Global *Pax Americana,* enforced by U.S. military might, is illogical and unachievable. However, by maintaining vigilance and revolutionary military capability, it will be difficult for anyone to seriously challenge us.[5] Isolationistic withdrawal from foreign intervention is equally illogical. The solution then becomes judicious application of force coupled with a redefinition of winning.

For any adversary to misread this concept and believe they could blatantly attack U.S. vital interests with impunity would be a perilous mistake. Any serious affront to our security or interests would be met with decisive, overwhelming force. In the immortal words of General Yamamoto following the Japanese Navy sneak attack on Pearl Harbor, they *would awaken a sleeping giant.* It is extremely unwise to get the American people really mad. Once the public becomes emotionally involved in an issue of vital interests, they will demand that their government take any and all steps necessary to resolve it—unconditionally. By nature, Americans tend to be slow to engage in international politics but quick to anger when core values are at stake. Once the line has been crossed, they are prepared to make whatever sacrifices are necessary to provide the resources required to crush any adversary that stands in their way.

Pragmatically, however, in the wide range of operations in which we will find ourselves in the future, it will be necessary to modify our thinking about how we define and measure objectives. Yes, that means that in some cases—like it or not—we must redefine the sacred concept called *winning.*

Throughout this book we have emphasized that there will continue to be threats to national security that demand conventional responses, primarily with highly lethal precision-guided weapons. The majority of our armed forces are rightfully structured to defeat any adversary foolish enough to confront our military in a traditional battle. Today, American armed forces are the best-trained and best-equipped soldiers in the world. That strength is further enhanced through strategic alliances that will be necessary to maintain regional stability around the globe.

However, in many ways the nature of conflict has changed. While we are prepared to defeat conventional foes, emerging threats—mercurial, surreptitious, and enigmatic—will prove to be more difficult to repress or constrain. For an increasing number of complex situations requiring a forceful response, traditional systems—bombs, missiles, and bullets—will be either ineffective or even counterproductive. To accomplish those limited-objective missions, part, but certainly not all, of the solution to the use of force in future conflicts lies in the rapid development and fielding of additional non-lethal weapons.

EPILOGUE

It's like monitoring a moving train.
—*General Hugh Shelton, Chairman of the Joint Chiefs of Staff, Senate Armed Services Hearings, 29 September 1998*

Any book that attempts to prognosticate future events in today's rapidly changing world is fraught with risks. Given the time necessary for the administrative aspects of publication, underlying situations can change dramatically or events may come to pass that make the predictions appear to be a blinding flash of the obvious rather than truly insightful.

So it has been in writing *Future War.* The bulk of the material you have read so far was submitted when the initial manuscript was completed in September 1997. When we agreed to delay release of the book until May 1999, I knew that an update would be required. The material presented is still valid. However, predicted events have occurred and there has been an increased awareness on the part of senior officials as well as the general public of the complex situations facing both the military and law enforcement.

From a non-lethal weapons perspective, the necessity for these systems has continued to be demonstrated all the way from individual restraints to strategic applications. Understanding of the problem by the military moves forward, albeit at a slow pace. The average plans and operations officer still has almost no comprehension of the weapons that might be made available to him or the advantages to be gained by their use. To better understand where we are today, let's begin in Africa.

Instability

"Base! Base! Terrorism! Terrorism!" As reported by *The Washington Post,* those were the words Benson Okuku Bwaku, a private security guard at the U.S. Embassy in Nairobi, Kenya, screamed out in a vain attempt to alert other guards of an impending truck bomb attack. Aware of his surroundings, Bwaku had noticed the suspicious truck sharply turn off Haile Selassie Avenue and accelerate towards his barricade. Blown off his feet moments after sounding the alarm, Bwaku could not have imagined that this incident would have global impact and change the face of terrorism.[1]

It was 10:40 A.M. on 7 August when the bomb in the Mitsubishi truck exploded, killing 257 people, mostly Africans, and injuring more than 5,000 others. The U.S. Embassy, as well as nearby buildings, were destroyed by the enormous concussion of the terrorist constructed bomb. Devastating? Yes, but we have been bombed before. It was the second terrorist event, occurring almost simultaneously at the U.S. Embassy in Dar es Salaam, capital of Tanzania, that signaled a new level of sophistication. The casualties were lower in Dar es Salaam: ten were dead and about seventy wounded, again mostly innocent African bystanders or embassy employees. However, it was the ability of the terrorists to closely coordinate the attacks that brought attention to these events.

Unlike investigations of other bombings, a picture of the terrorists quickly emerged. Within a very short time the location of the bomb assembly had been identified and several suspects were in custody in Kenya and Pakistan. It was immediately assumed, and later proven to be true, that the one who had orchestrated the attacks was Osama Bin Laden, the same Saudi dissident mentioned earlier in Chapter 2. Among those arrested was Mohammed Saddiq Odeh, a Palestinian from Jordan. Odeh, a known associate of Bin Laden, was spotted when he attempted to enter Karachi on a forged passport. Not only did he confess to helping plan the Nairobi attack, but also to training some of the Somalis involved in the attack on the U.S. Rangers in Mogadishu. Captured in Kenya was Mohammed Rashed Daoud al Owhali. He had been a passenger in the truck, and the planners expected that he would die in the bombing attack. During FBI interrogation, Owhali admitted he had been trained at Bin Laden's training bases located deep in Afghanistan.[2]

Once known only to a few intelligence analysts, Bin Laden instantly became a household name. Reports of his activities filled newspapers, magazines, and television programs. Rarely seen, Bin Laden enjoys the protection of the Taliban in Afghanistan. From there he has called for a *jihad,* a holy war, against the United States, even stating that, in addition to the military, civilians were fair targets. In an interview with ABC News before the bombing, the tragic events were foretold. In June 1998 Bin Laden told ABC correspondent John Miller that his Islamic terrorists would "send the bodies of American troops and civilians home in wooden boxes and coffins. We don't differentiate between those dressed in military uniforms and civilians. They are all targets in this fatwa."[3] Through the vicious acts in Nairobi and Dar es Salaam, Bin Laden also made it clear he had no regard for the lives of innocent bystanders who were not on his approved target list: in this case, Kenyans and Tanzanians.

Provided with hard information that Bin Laden was directly involved in supporting the terrorist attacks, the United States launched a substantial attack against his bases in Afghanistan. While a ground operation to snatch Bin Laden had been contemplated for several months, the difficulty in reaching into the distant rugged terrain rendered that option nearly suicidal. Therefore, air attacks were planned. The logical choice of weapons was cruise missiles. They had long range and could strike with great precision. Pilotless, there would be zero chance that an airman might be downed in the forbidding mountains that the camps occupied. Planners were well aware of the merciless torture that had befallen Soviet troops unfortunate enough to be captured by mujahedeen who viewed human suffering as a spectator sport.

Very reliable intelligence sources learned that Bin Laden had called a meeting of many of his top lieutenants at a training base near Khost. Ironically, these bases were built partially with U.S. funding provided to the mujahedeen by the CIA when the United States was backing their efforts against the Soviets. Then, the Afghan resistance had received nearly $6 billion from U.S. and Saudi sources.[4] The six training camps of the Khost complex were easily defended from ground assaults and designed to withstand extensive air strikes. The Soviets had attacked the area with their elite Spetznaz special operations units on several occasions. Each time they suffered devastating defeats, sometimes with over 50 percent of the unit killed. According to Milt Beardon, the retired CIA operations chief who had supported the mujahedeen for years, merely striking the camps with bombs would "just move the rubble around." He did note that going after the people might be worthwhile.[5]

Two targets were selected for cruise-missile attack. One was the meeting place selected by Bin Laden, the Zhawar Kili Al-Badr training camp located against the mountains about one mile from the Pakistani border. The second was on another continent, the Shifa Pharmaceutical Plant near Khartoum, the capital of Sudan. According to intelligence sources, the Shifa plant was supporting Bin Laden and engaged in making precursor chemicals for the nerve agent XV.[6]

At 10 P.M. in Afghanistan on 20 August, Operation Infinite Reach commenced with a barrage of sixty Tomahawk missiles fired from U.S. Navy warships in the Arabian Sea off the shores of Pakistan. At 7:30 P.M. local time in Sudan, up to twenty cruise missiles launched from two naval ships in the Red Sea against the Shifa plant impacted. Evening had been designated for the attack that was intent on destroying the factory while minimizing the chance of killing workers or bystanders.[7, 8]

Within hours reconnaissance systems revealed that heavy structural damage had been done at both target sites. However, Bin Laden emerged unscathed and very angry. Soon he stated that he had been warned about the impending attack and had canceled the key meeting in time to prevent any of his senior staff from being killed. This demonstrated the problems associated with keeping operations secret. Inside the beltway only a select few, known as the Small Group, were aware of the operation. Yet somehow, it appears that the information found its way to Bin Laden in time for him to respond.[9]

While the senior officials said killing him was not the objective of the raid, everyone else assumed it was. Bin Laden was quick to respond. After he announced his survival his support groups began declaring, "The war has begun." Faxes from the International Islamic Front for Holy War Against Jews and Crusaders pledged to mount "pitiless and violent attacks against Israeli and U.S. targets." Citing the London *Times,* the article also noted the level of sophistication terrorists now includes employing business corporations as fronts to launder money and provide cover for the terrorists.[10]

From Khartoum a great uproar arose claiming the factory was only engaged in making legitimate pharmaceutical products, ones desperately needed in the area. The U.S. officials countered that there was hard evidence that Al Shifa was making the agent for VX and had ties to Bin Laden. It was also suspected that the plant had ties to Iraq and was used to thwart efforts of UN inspectors looking for weapons of mass destruction internally. According to senior U.S. intelligence officials, CIA had obtained a soil sample taken just outside the plant and found it contained traces of ethyl methylphosphonothionate, known commonly as Empta, a substance then claimed to have no other use than to make VX nerve agent. Further, the plant manager, Osman Sulayman, was deported from Saudi Arabia because of ties with Bin Laden.[11] Noting a long history of support of terrorists, in 1996 Madeleine Albright had called Sudan "a viper's nest." Sudan sheltered Egyptian fundamentalists who attempted to assassinate President Hosni Mubarak and was a known base for Bin Laden operatives.

Within days, however, despite the evidence cited, some Department of State and intelligence community sources were breaking ranks. While Sudan certainly did support terrorists, the evidence linking the production of VX precursor to Al Shifa seemed weaker than expected. Some stated the evidence was not sufficient to warrant such a retaliatory strike.[12] Additionally, many felt that President Clinton had generated a "Wag the Dog" scenario to distract attention from his mounting personal problems. The concept was based on a 1998 movie starring Robert De Niro and Dustin Hoffman in which a politically embattled U.S. President creates the illusion of a war in order to win his reelection.[13]

The raid against the Shifa plant near Khartoum directly relates to the strategic implications of non-lethal weapons. Located near a densely populated area, the chance for collateral casualties was fairly high. In this case, the plant did not operate at night, and all of the missiles hit the target.

Such a favorable situation cannot be guaranteed in future conflicts. It has been noted that targets located in cities, possibly close to sensitive places such as hospitals, schools, or religious buildings must be anticipated. High explosive warheads will not always be acceptable. Alternatives must be found. For instance, chemical and biological plants require extensive air conditioning or filtration systems. The filter systems are vulnerable to several non-lethal agents that can be used to cause production to shut down for long periods of time. Other critical pieces of machinery can be degraded so as to make the facility inoperative. Power systems can be taken off-line and buildings be made uninhabitable for the work force. What is needed is the development of systems capable of results as an alternative to current kinetic solutions. These are necessary strategic options for the complex situations of the future.

The terrorist attacks in Africa were predictable. While no one could foretell the precise time or location, they had to happen. In general, regional stability is decreasing around the world, and the outlook is very messy. The global economy has taken a severe beating. Investors have watched once booming markets plunge, currencies be devalued, and psychological confidence erode. The Russian economy experienced a meltdown that also impacted many foreign investors. Devaluation of the baht in Thailand began a trend of currency manipulation that spread throughout Asia. Countries like Japan, Brazil, and Germany, which once led regional economic prosperity, have suffered substantial setbacks that have had a domino effect on both regional and global economic security. The interconnectedness of regional economies, all demonstrating profound instability, raised concerns of a global collapse. Efforts by the G-7 nations to establish international policy were indecisive. As stated by Greek Finance Minister Yannos Papantoniou, there was a "lack of clear political will to move in any direction."[14]

The economic downturn has been coupled with politically weakened leadership in several major countries, including the United States and Russia, and with no international consensus on how to reestablish stability. Concurrently, several smaller countries have experienced domestic turmoil. In Indonesia, the world's fourth most populous country, their long-term leader, Suharto, was forced out after a period of rioting. Following that, Malaysia was in political upheaval. North Korea seemed internally stable but was on the verge of mass starvation. In southern Asia the Taliban in Afghanistan threatened Iran, while perennial hostilities continued between the newly declared nuclear states of Pakistan and India. In Africa, Algerian massacres became endemic, with thousands of people being hacked to death. Congo reverted back to civil war, with six armies attempting to occupy territory. Angola is on the verge of civil war against UNITA, which will likely bring Zambia into the conflict. Rwanda continues to be problematic. Oil-rich Nigerian leaders fail to improve the abject poverty of native Ogoni people, who are close to rioting. Religiously motivated civil war in Sudan has created dramatic food shortages in the south. Famine takes a heavy toll in Ethiopia, which still somehow finds the means to wage war with Eritrea. The unrest in Africa is fueled by large weapons sales from Bulgaria and France.

On America's southern border, corruption within the Mexican government sponsored by endless supplies of drugs and illegal money reached an intolerable level. While the Cali and Medellín cartels were structurally damaged through concerted efforts of law enforcement agencies, others have risen to fill the gaps and Mexico has suffered. Additionally, there is constant tension in the southern state of Chiapas between the indigenous Indian population and the central government. Beyond Mexico, Dominicans have now stepped into the drug trade and are considered the dominant drug lords on the East Coast of the U.S. While the American Special Forces and SEALs assist Peru in fighting the drug wars, Colombian rebels have those resisted efforts in the state of Guaviare where 65 percent of all of the cocaine is grown.

The Balkan states remain unpredictable. Peace support operations have had to be extended well beyond the dates originally agreed upon. Periodic bombing, or threats of bombing seem necessary to keep any semblance of civility. The extent to which military support might be required seems indefinable.

This high degree of instability at flash points around the world has a direct bearing on the development and deployment of non-lethal weapons. It seems clear that use of force will be required to resolve, however temporarily, disputes in many areas. Peace support operations and humanitarian missions are likely to increase. Keeping factions bent on conflict apart will be deemed in the best interests of all countries. Economic strife will invariably lead to both domestic and international discord. Therefore, having force options that minimize the potential for lethal consequences are absolutely essential.

Discussed earlier were issues relating to the impact humanitarian and peace support operations would have on troop retention and strength. The Joint Chiefs of Staff finally acknowledged that the cuts made in military strength have reached a critical level. In my opinion, we went below a minimum safe level some time ago. As Army Chief of Staff General Reimer pointed out in Senate hearings, only twice before in history have we drawn down to such dangerously low levels. Both led to war. As he put it, "We have too few resources chasing too many requirements."[15] Democratic Senator John Glenn noted that strengths have been cut from a high of approximately 2.1 million in uniform to a current level of 1.4 million. In his view, the minimum safe level was 1.6 million.[16]

Unlike the other service chiefs, Marine Corps Commandant General Charles Krulak has been sounding the alarm for some time. In those hearings he noted that we cannot have a "911 force and modernize" with the current budget.[17] He was referring to the dilemma faced in being sent out frequently on unanticipated peace-support missions, told to pay for the missions from existing funds, and at the same time develop and field advanced new weapons systems.

The development of non-lethal weapons is caught up in this dilemma. While there has been some movement on the part of the Joint Non-Lethal Weapons Directorate, it is slow and modest. While they have made gains in policy, their budget remains very limited, and is unlikely to grow without external impetus. It will be necessary for visionary commanders to comprehend the advantages to be gained by having a comprehensive arsenal of non-lethal weapons, then establishing their requirements and fencing the funds.

Training for use of non-lethal weapons has improved significantly. Both the Army and Marine Corps have been involved in creating a training course so that soldiers will be better prepared to use these systems when the occasion arises. A big step forward was taken on 19 June 1998 when the services approved the Non-Lethal Individuals Weapons Instructor Course. Captain Steve Simpson and Gunny Sergeant Steve Carlson, Marines stationed at the Army Military Police School, developed the program so that soldiers around the world could receive this vital training.[18]

High-level support for non-lethal weapons came in the summer of 1998 when Bob Bell, Special Assistant to the President for National Security, sent a letter to Deputy Secretary of Defense John Hamre directing a study be conducted to determine if a national policy should be developed. The study incorporates the knowledge and experience of some of the most senior policy makers and advisers. Support from ex-cabinet level policy makers could help move the visibility of the issues related to non-lethal weapons from relative obscurity to the front burner.

Important articles continue to be published in support of non-lethal weapons but more are needed. The strategic implications for the Air Force were grasped by Colonel Joseph Siniscalchi and detailed in an occasional paper for Air University. He concludes, "The larger radius of effects can enable a near instantaneous attack on numerous, virtually unlimited, strategic centers of gravity." Colonel Siniscalchi addresses the value of investing in research and development of non-lethal systems.[19] His paper is important as full integration of non-lethal weapons into the American arsenal will only occur after the strategic and operational advantages are better recognized. Articles by other senior officers are needed to assist in the cumbersome educational process.

Non-lethal weapons have been incorporated into some high-level wargames, but only as adjuncts or excursions. In April 1998, lead by Ed Scannell of the Army Research Laboratory, a number of us participated in the Army After Next wargame at Carlisle Barracks. The Army After Next is a massive development process to design ground forces for the future decades. The good news is that we were included. The bad news is that we had to work outside the main game, and could not be integrated into their play. That was unfortunate, as the blue force scenario could have benefited significantly from non-lethal weapons. The problem was simply that the operations officers and commanders in the main game had insufficient knowledge of what could be gained from non-lethal weapons.

A similar experience was found at the annual naval wargame, *Global 98,* conducted at the Naval War College in the summer of 1998. Again it was reported that blue forces did not employ non-lethal weapons in the main game.

However, in an excursion, red forces were able to stop blue using non-lethal systems. In all services there is a lack of understanding of non-lethal weapons capabilities.

Interestingly, the Russians reported learning some of the similar lessons about non-lethal weapons in their urban war with the Chechens in Grozny. In their after-action report, the Russians noted that they had underestimated the importance of cultural differences and lost the information war that the Chechens conducted with cell phones, faxes, and other modern communications means. They needed the capability to separate combatants from innocent civilians and the requirements for anti-personnel non-lethal weapons.

Concern about the activities of militias from within the United States continues to grow. In a study conducted at the University of Florida, Keith Akins found militia members to be better educated than the general population, and armed beyond the expectations of most law enforcement agencies. Religion seems to play a role. Those people who are members of the estimated 440 militia groups tend to be church members and often see a failure of governments to "institute Christian law." Akins did note there was a substantial danger to innocent people who might be caught by a random bombing.[20] It is in responding to militia threats involving standoffs that non-lethal weapons will play an important role.

Law enforcement also continues its search for additional non-lethal weapons. They are becoming increasingly alarmed at the number of incidents indicative of "suicide by cop." The problem appears to be far greater than previously supposed. The Los Angeles Sheriff's Office conducted a study of the shooting deaths by deputies over the past decade. They concluded that nearly 10 percent of such shootings was intentionally provoked by the victim in order to bring about his, or occasionally her, own death. In most cases they actually ask the police to kill them. To meet this need new tactics and negotiation strategies are being developed. So are non-lethal weapons.[21]

Infrastructure

Among the biggest changes to occur in the intervening year between submitting the draft manuscript and this update has been the formal recognition of the vulnerability of the U.S. national infrastructure to terrorist attacks. The panels being formed a year ago have since met and begun to take action. On 22 May 1998, President Clinton announced formation of a new effort to defeat terrorism and threats to our national infrastructure. In his commencement speech at the U.S. Naval Academy in Annapolis, Maryland, he stated he was appointing a National Coordinator for Security, Infrastructure Protection, and Counterterrorism. This was in recognition of the scope of the problem and to ensure that diverse agencies would cooperate with each other. He named Richard Clarke, current Special Assistant to the President for Global Affairs, to head the operation.[22]

Recently, information about two Presidential Decision Directives that have impact on our infrastructure have been made public. In Congressional hearings, Dr. Jeffrey Hunker, Director of the Critical Infrastructure Assurance Office, described what had been accomplished. The first directive, PDD-62, signed in 1996 contained major initiatives to combat international terrorism. The second, PDD-63 addressed protection of the nation's critical infrastructure from both physical and cyber attacks. PDD-63 identifies critical infrastructure protection as a national security priority. The plans call for vulnerability assessment, remediation of vulnerabilities, a national warning system, and responses to attacks in progress. Additionally, they will sponsor education and awareness programs for individuals and propose to support research and development of technologies to minimize vulnerabilities.[23] In order to respond to attacks generated inside the United States, non-lethal weapons should be part of that package. They may be useful in other venues as well.

There have been highly publicized exercises to train those who are likley to be called upon in the event of a major incident such as an attack with chemical or biological weapons. It has been acknowledged that it will not be specialized federal forces that become the first line of defense against such attacks. Rather it will be local "first responders" such as police, firefighters, or ambulance personnel, who encounter the threats and sustain great personal risk in assisting victims. Training these people who may be faced with very ambiguous, yet terrifying, situations is critical.

The extent of the threat was described in Congressional Hearings before the Joint Economic Committee chaired by Congressman Jim Saxton. One of the most alarming speakers was Dr. Ken Alibek, the previously mentioned former First Deputy Director of Biopreparat, the civilian arm of the Russian biological weapons program. In his presentation he provided more information than previously has been made public about the extent of their program and the dangers involved. The potential biological agents have been described earlier. However, Alibek also provided information regarding the economic impact of a BW attack based on a Centers for Disease Control model run in 1997. According to predictions, for each 100,000 persons exposed, the impact to the U.S. economy would range from $477.7 million to $26.2 billion.[24] Alibek told me that the Russians had concentrated on development of antipersonnel weapons. He did not believe they had seriously investigated antimatérial systems.[25] In my opinion, these antimatérial weapons are still a significant threat to our critical infrastructure.

Another class of weapons has emerged as a far greater danger than previously supposed. Pulsed power systems were discussed in the chapter on electromagnetic weapons. One category, radio-frequency (RF) weapons, have risen in concern to the level that a formal national intelligence estimate (NIE) has been coordinated by the National Intelligence Council. I had the opportunity to participate in several of the meetings and hear firsthand the potential damage that could be inflicted against unprotected electronics such as most computer systems.

These weapons were also described for the Congressional Joint Economic Council. One speaker was Victor Sheymov, a former KGB major who defected to the West years ago. His KGB assignment in the Eighth Chief Directorate involved protecting Soviet classified transmissions between embassies around the world and acquainted him with various technical aspects of systems vulnerabilities. His defection, the harrowing tale of which is told in his book *Tower of Secrets,* was a coup for U.S. intelligence.[26] At the hearings he described the effects of low-energy RF weapons that could be "devastating and highly indiscriminate." He noted the development of these weapons does not require high technology. Sheymov noted a shoebox-sized weapon could be constructed in less than three hours using store-bought electrical components.[27] In private conversations he indicated a great depth of knowledge as to how sophisticated weapons could be employed against civilian targets. Properly done, he suggested, entire buildings could be subjected to repeated electrical-supply malfunctions, without the source of the real problem ever being discovered. [28] Clearly, these non-lethal RF weapons could play a major role in an attack on our infrastructure. Conversely, they also offer interesting offensive capabilities that can be used to degrade a potential enemy or to send a message of our resolve.

American awareness of threats from computer hackers increased significantly in 1998. As previously stated, information warfare, while non-lethal in nature, has become so extensive as to be regarded as a separate category. At the national level, planners and analysts should view information warfare as part and parcel of strategic non-lethal concepts.

As a result of early 1998 intrusions into various Defense computer systems, Deputy Secretary of Defense John Hamre established a permanent joint task force on cybersecurity.[29] There are some efforts to coordinate cyberprotection mechanisms to the civilian information infrastructure as well. To gain an appreciation of how hackers attack and of reasonable security methods, please see the series of articles in the October 1998 issue of *Scientific American.*[30,31,32] It is fair to say that concentrated hacker attacks against civilian computer systems in communications, transportation, finance, or other information dependent businessess could seriously jeopardize our national security. Measures are finally under way to increase security across the board.

Worthy of note are the writings of Chinese senior military officers. Contained in most articles addressing future conflicts is the importance of information warfare. They have read our publicly available doctrine and are making plans accordingly. Writers such as Major General Wang Pufeng, former Director of the Strategy Department of the Academy of Military Science in Beijing, talks of developing a new guide to action in which "information warfare theory is new warfare theory" and should be used to advance China's capabilities. These military officers advocate targeting both our military command and control systems and our critical infrastructure.[33]

Conclusion

Current military planning, in the words of the Chairman of the JCS, General Hugh Shelton, is "like monitoring a moving train." The same is true for getting a firm fix on the state of non-lethal weapons developments and requirements. Administratively, there have been modest advances on non-lethal weapons in the past year. These have been due to two factors. One is the hard work of Colonel Andy Mazzara and his staff at the Joint Non-Lethal Weapons Directorate and their associates. The second is support from senior officials who are prepared to push for new options.

NATO has also moved ahead in developing international policy. That group is led by the American delegate, once again demonstrating that the Europeans are willing to let the United States set the pace with development of new non-lethal weapons. Of course, many of them already have more experience with controlling civil disobedience and terrorism than we have. Continued extensive involvement in peace support operations such as Bosnia will undoubtedly increase the demand for more and better non-lethal weapons.

American vulnerabilities lie within ourselves. We have grown more sensitive to all casualties, not just those of our troops. This limits our ability to exercise the full extent of power available to us.[34] Addressing the United Nations on 21 September 1998, shortly after the bombings of our embassies in Africa, President Clinton noted that terrorists are taking advantage of the openness of democratic societies and that this is a threat to all mankind. He noted when it comes to terrorism there "should be no dividing line between Muslims and Jews, Protestants and Catholics, Serbs and Albanians, developed societies and emerging nations."[35]

As predicted, the world is becoming increasingly unstable. In many, if not most, of the potential areas for conflict, use of overwhelming force is unwarranted, unwise, or impractical. The same is true in countering terrorism or in apprehension of well-armed criminals. We have allowed our military strength to dwindle to a dangerous level at a time when foreign commitments are increasing.We are not asking them to defeat an adversary, but rather help maintain stability in very dangerous situations. We have an obligation to ensure they have the tools necessary—including a substantially upgraded non-lethal weapons arsenal.

Appendix A

A MODEL FOR UNDERSTANDING SOCIAL ORGANIZATIONS

The basis for future societal organization may be described by examining three factors. They are structure, resources, and values as depicted on the axes in Figure 1. Each axis addresses one component. The vertical axis describes the possible institutional structures from large groupings to smaller ones. Any given organization will have some form of structure that identifies it. Some of the structures may be different from what we have previously experienced in social organizations. The horizontal axis depicts the resource categories that may be available to those organizations. It is critical to know what resources are available to an organization at any given time. And the diagonal axis represents the institutional values that are held by each organization. Organizations will allocate resources based on their values and beliefs. It is a common mistake to assume that there is a single logical method by which societies will organize. That is the "rational man" approach based on traditional Western values. It has been demonstrated repeatedly that foreign countries do not always respond as we predict. Though frequently labeled "crazy," it is more likely that they have applied different values in arriving at their solutions to problems.

While many have spoken to these subjects, this model for thinking about social organization has not been previously available. It provides a method whereby analysts can more accurately predict the courses of action that potential adversaries may take.

Structures. The possible structures run from the currently dominant unit, the *nation-state,* to commonly accepted supernational and subnational entities, but may include emerging concepts such as virtual organizations. *Supernational* organizations would include the European Union, NAFTA, and will probably see the rise of a Pacific Rim economic bloc in the near future. *Transnational* organizations have previously been addressed but can include legitimate economic entities with sufficient wealth to be able to become players in regional affairs. *Subnational* elements include religious and ethnic groups that are normally collocated with others. Some come from "failed states," while others are the product of previous geographic subdivisions. Too frequently, prior geographic boundaries were drawn based on conquest or cartographers' whims

The Basis of Societal Structures

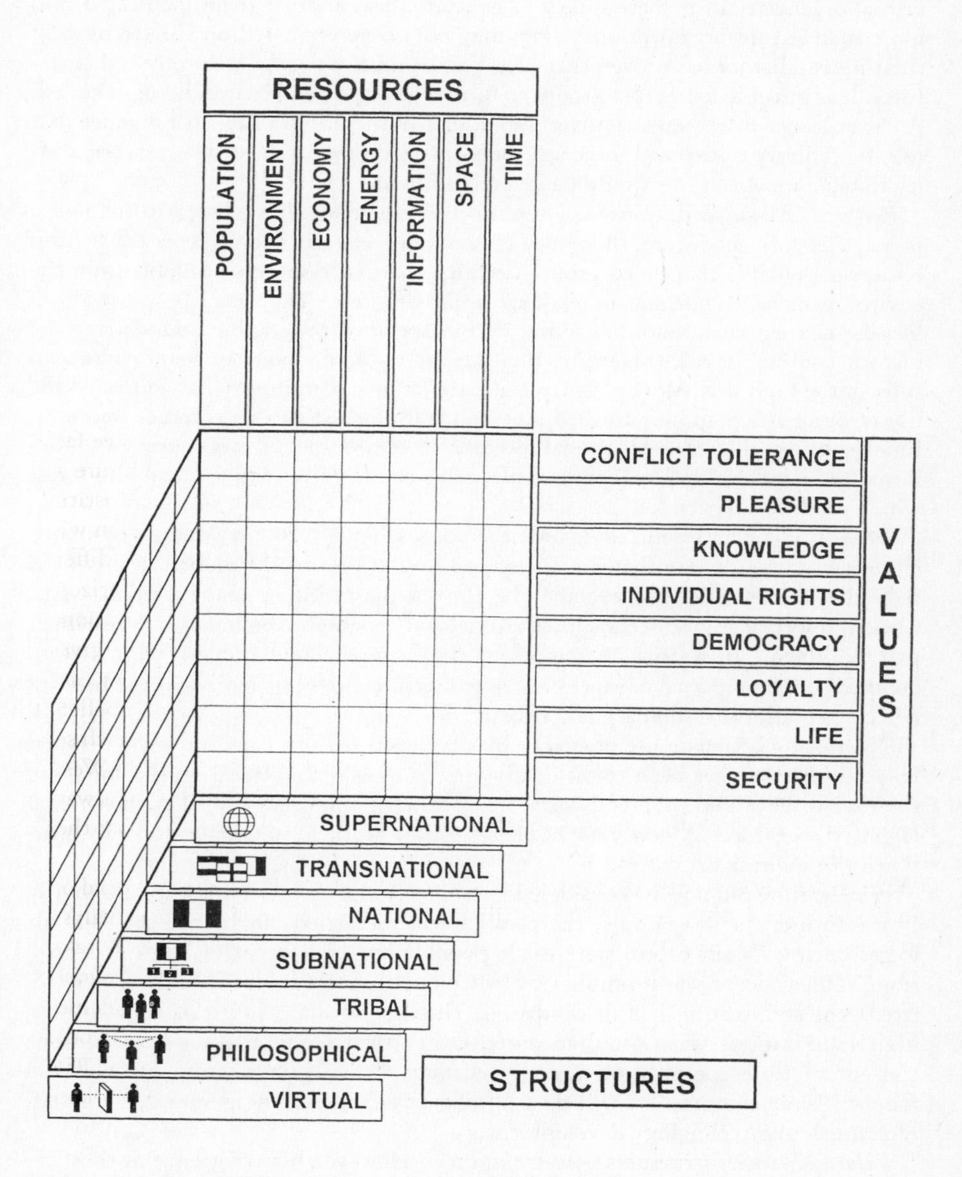

and did not properly account for the demographics or sentiments of the indigenous population.

Virtual organizations are ones that usually have no contiguous physical structure, and with whom affiliation is based on an ideological and voluntary basis. Burgeoning information technology will enable such groups to form, conduct business, and disestablish, leaving little trace of their physical existence. The importance of such virtual organizations is their ability to generate ideas and get them inculcated into more mainstream organizations. They may obtain power based on the strength of their ideas. History has proven that ideas can be more powerful than physical structures. It is possible for virtual groups to form without true identities being revealed. A clever leader may then stimulate individuals of the group to act in a manner that may be contrary to national interests, then simply disappear. In such a case, our ability to retaliate with force would be severely limited.

Resources. All organizations have resources, some of which are listed in the figure on page 219. In the future, there may be some new concepts about what constitutes resources available to a social group. Certainly natural resources available from the environment will continue to play an important role. They may be placed in a broader context than materials found within specified geographic boundaries. The historic context for rules regarding these resources can be found in water rights and rules of the high seas. More recently, the international community has addressed the use of outer space and established accepted rules for use of this common resource. Emerging trends hold international interests in environmental issues previously believed to be internal affairs. For example, can a country decimate its own rain forest which produces oxygen for the globe?

Other traditional resources include the population of a country and its economy. These concepts, too, are changing, based on migration and the globalization of the economy. Information has become the coin-of-the-realm in many organizations. Clearly, information transcends institutional and geographic boundaries in a manner that is difficult to restrict. Attempts to codify intellectual property rights seem doomed to the perpetual advances in related technologies but remain at the heart of fierce international diplomacy and debate.

The disproportionate use of energy by developed nations is an intensely debated issue. Energy use has been traditionally viewed as available based on the ability of the user to pay. The collateral costs, i.e., contamination, have until recently been largely disregarded. A new view of energy rights is likely to emerge and add complexity to societal structures.

Finally, time can now be considered a resource and may be viewed very differently in the future. As an example, the parallel attacks carried out by the coalition air forces during Desert Storm were made possible by the concept of "time compression." This concept was promulgated by Colonel John Warden, U.S. Air Force (retired), the architect of that air campaign. Through advances in precision bombing, high-value targets were simultaneously, rather than sequentially, destroyed. The concept of time compression literally changes the way battles are planned and fought.[1] Time compression will be a significant factor in areas as varied as combat, economics, and technology development.

Values. All social structures possess resources. However, how they employ those resources varies, based on the prevailing value system. Some of the values will be held at a fundamental level, thus are relatively firm. Others will be transient and based on a hierarchy of need, similar to that proposed by psychologist Abraham Maslow for individuals.[2]

At the most basic level of values, societies, like individuals, perceive a need for physical security. Thus, they have almost invariably raised and maintained armed forces. While they value security highly, the definition of security may vary greatly from one societal organization to another. One structure will remain purely defensive and within designated boundaries, while another may view preemptive strikes or invasion of foreign territory as a natural extension of their security requirements.

Another area of contention is the value placed on life, human or animal. Advanced societies generally accept that human life has a relatively high value, certainly superior to material objects. Other societies have placed lower value on human existence. Sometimes fostered or rationalized by religion, they employ their population resources in modes very different from other societal groups. Clashes are almost inevitable when such deep-seated value differentials are encountered. Today, there is heated political debate about the rights and obligations of external parties to intervene when human rights are at stake.[3]

Social values are not well understood. The information age infuses ideas and concepts at an unprecedented rate. It is likely that an individual can belong to several social organizations or groups and that those groups may have competing value systems. Individuals will then have to choose which values they will embrace at any given time. Unfortunately, this makes people's behavior less predictable. Pragmatic social entities will manipulate belief and value systems to their advantage, as is evident in American politics. The end result is a transient value system yielding changing allegiance.

Appendix B

NON-LETHAL WEAPONS TAXONOMY

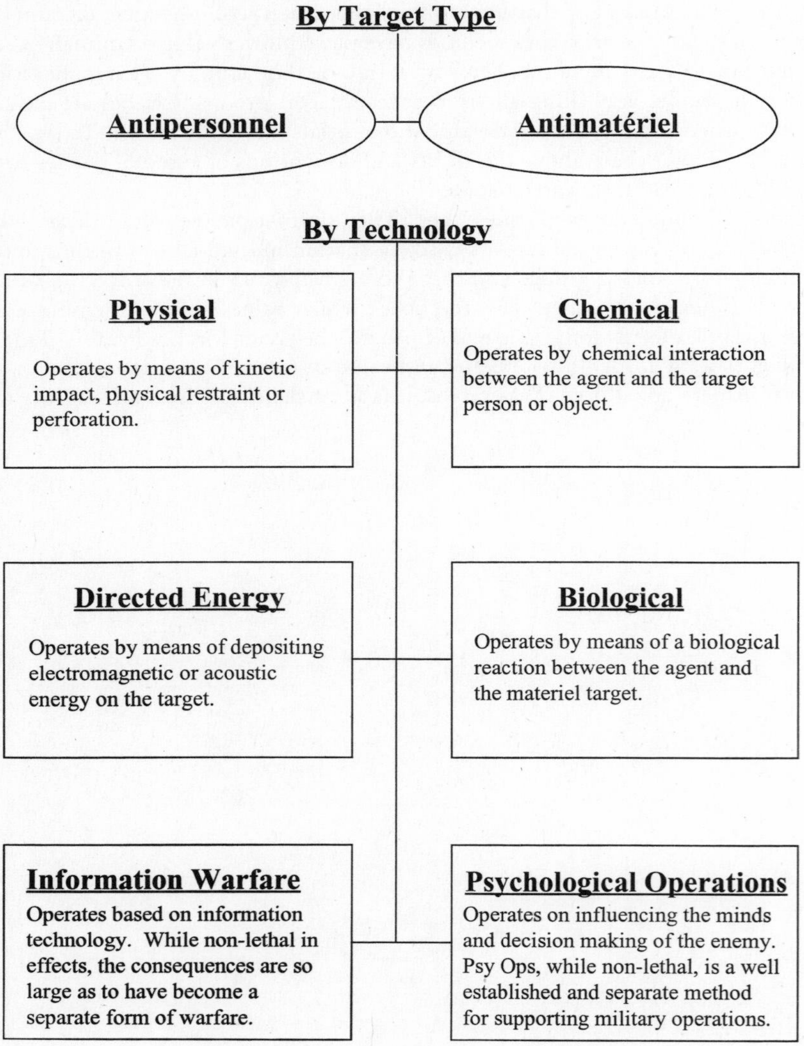

By Target Type

Antipersonnel

Antimatériel

By Technology

Physical

Operates by means of kinetic impact, physical restraint or perforation.

Chemical

Operates by chemical interaction between the agent and the target person or object.

Directed Energy

Operates by means of depositing electromagnetic or acoustic energy on the target.

Biological

Operates by means of a biological reaction between the agent and the materiel target.

Information Warfare

Operates based on information technology. While non-lethal in effects, the consequences are so large as to have become a separate form of warfare.

Psychological Operations

Operates on influencing the minds and decision making of the enemy. Psy Ops, while non-lethal, is a well established and separate method for supporting military operations.

ANTI-PERSONNEL NON-LETHAL WEAPONS

Physical
- Rubber/plastic Rounds
- Beanbag Rounds
- Ring Airfoil Grenades
- Wooden Batons
- Foam Batons
- Rubber Pellets
- Water Cannon
- Nets
- Compressed Air

Chemical
- Irritants (RCAs)
  - CN/CS
  - Pepper spray (OC)
- Dyes
- Olfactory Agents
- Sticky Foams
- Nauseating Agents
- Calmatives
- Hallucinogens
- Obscurants

Directed Energy
- Electromagnetic
  - Dazzling Light
  - Flash-Bang Grenade
  - Eye-safe Laser
  - Stun Gun
  - Pulsing Lights
  - Microwaves
  - Holograms
- Acoustic
  - Loud Audible
  - Infrasound

Biological

No Legal Anti-personnel Agents

Terrorists or others may use Pathogens

ANTIMATÉRIAL NON-LETHAL WEAPONS

Physical
- Vehicle Nets
- Wire Entanglements
- Fiber Entanglements
- Caltrops
- Spike Strips

Chemical
- Combustion Modification
- Filter Cloggers
- Sticky Foams
- Viscosification Agents
- Obscurants
- Aggressive Agents
  - Superacids
  - Caustics
  - Solvating
  - Embrittlement
  - Friction Enhancers
  - Catalytic Deploymers
- Metal Fibers/Particles
- Environmental Control
- Friction Reduction

Directed Energy
- Electromagnetic
  - Pulsed Power
  - High-Power Microwave
  - Direct Injection
  - NNEMP
  - Counter Sensor Laser
  - Jamming
  - Particle Beams
- Acoustic
  - Infrasound
  - Ultrasound

Biological
- Bioremediation Agents
  - Petroleum Products
  - Metals
  - Plasticizers
  - Concrete
  - Explosives

TARGET CATEGORIES FOR NON-LETHAL WEAPONS

People

- Combatants
- Criminals
- Hostages
- Hostages (willing)
- Non-combatants
- Rioters
- Refugees
- Disaster Victims

Weapon Systems

- Optics
- Sensors
- Vision Obscurants
- Electronics
- Engines
- Fuels Supplies
- Tires
- Trafficability
- Ammunition
- Structural Fatigue
- Computer Programs

Infrastructure

- Telecommunications
  - Radio & TV Stations
  - Telephones
  - Computers
  - Satellite Links
- Transportation
  - Fuel
  - Tires
  - Roads
  - Bridges
  - Runways
  - Air Traffic Control
  - Shipping
- Energy System
  - Power Grids
  - Generators
  - Fuel Reserves
- Finance System
- Manufacturing
- Popular Support

SPECTRUM OF USE OF NON-LETHAL WEAPONS

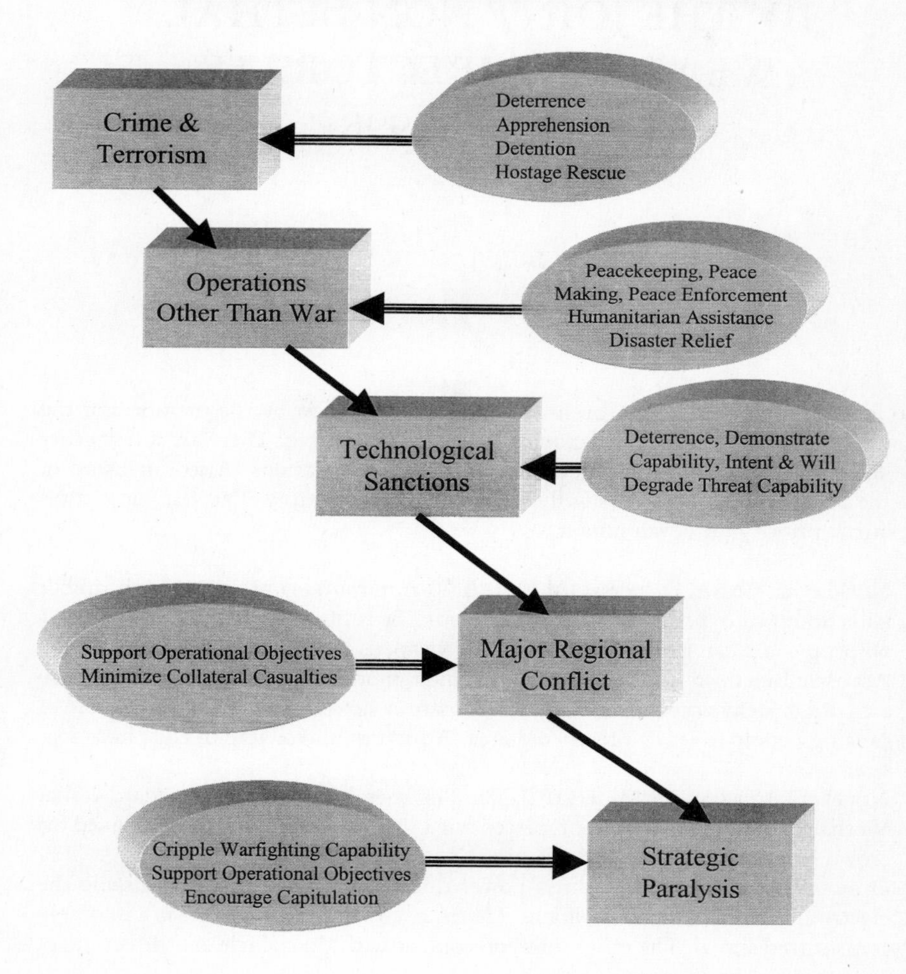

Appendix C

CURRENT PROGRAMS SUPPORTED BY THE JOINT NON-LETHAL WEAPONS DIRECTORATE (As of Fall 1998)

The following programs have been reviewed and agreed to by the members of the Joint Non-Lethal Weapons Directorate Steering Committee. There are a few other non-lethal systems being developed under other organizations. Also, inclusion of specific programs and priorities will change from time to time. This list constitutes the initial priority for development.

1. Non-Lethal Crowd Dispersal (M203). A 40-mm round containing small rubber balls designed to break up an unruly crowd. The rubber balls cause blunt impact on people and can be fired from standard weapons.
2. Acoustic Bioeffects. A crew-served or vehicle-mounted weapon used to disperse a crowd or deny access to an area. It generates a pressure wave that is capable of causing people to leave a restricted area. A picture is available in the photo section.
3. Non-Lethal Claymore Mine (MCCM). The metal balls of the standard lethal Vietnam vintage claymore are replaced with small rubber balls. It is designed for crowd dispersal and vehicle protection.
4. Ground Vehicle Stopper. A high-power microwave device that can disable the electronic components of a vehicle. There are two approaches. One is a remotely transmitted signal. The other relies on contact with the vehicle and direct injection of the pulsed power.
5. Maritime Vessel Stoppers. The intent is to develop a device that can stop inboard diesel engines of surface vessels without injury to the occupants. Specific technologies have not been identified.
6. Classifed Program (Antipersonnel).
7. Speed Bump (Net). Another vehicle stopper, this device employs a net with vinyl webbing strong enough to capture a speeding car. It is designed to snare a 5,100-pound vehicle traveling at speeds up to sixty mph and stop it within 200 feet.

8. 66-mm Vehicle-Launched Payload. This is a flash-bang device designed to deter crowds from attacking vehicles.
9. Unattended Aerial Vehicle (UAV) Non-Lethal Payloads. This system is designated to take advantage of the emerging UAV platforms. A variety of payloads is being studied. UAVs have the advantage of delivering NL munitions without placing soldiers at risk.
10. Bounding Non-Lethal Munition. The first of non-lethal antipersonnel mines, this employs existing mines but deploys a net instead of an explosive. The bounding action comes from the "Bouncing Betty" or VOLCANO mines that project their payload into the air.
11. Canister-Launched Area Denial System (CLADS). This system, mounted on a HMMWV, is designed to launch up to 20 canisters carrying various non-lethal payloads. Some of the payloads being considered include concertina wire, bounding nets, and malodorous substances.
12. Foam Applications. This is designed as an area denial system employing rigid foams and epoxies. The fast curing foams can quickly seal doors and windows or fill designated areas.
13. Vortex Ring Gun. This weapon applies gas impulses with flash, concussion, and non-lethal agents. It is designed to provide crowd control and area denial. Marking agents can be projected to allow later identification of people at a demonstration. It is a retrofit for an existing MK19-3 40-mm grenade launcher allowing for quick transition to lethal capability if necessary.
14. Underbarrel Tactical Payload Delivery System. This is designed to attach under a standard M16 rifle. The "under" portion operates using compressed air to project various non-lethal payloads such as dyes, OC, or kinetic impact rounds. The rifle can instantly be used in standard lethal mode when the situation requires.

NOTES

1. *Why Are We Doing This?*

1. Joseph F. Coates, *Nonlethal and Nondestructive Combat in Cities Overseas,* Institute for Defense Analysis, Science and Technology Division, Washington, D.C., 1970.
2. Karl von Clausewitz, *On War,* translated by J. J. Graham (London, 1908).
3. Sun Tzu, *The Art of War,* translated by Thomas Cleary (Shambhala Publications, 1988).
4. Colonel John B. Alexander, "Anti-materiel Technology," *Military Review,* vol. LXIX, no. 10, October 1989.
5. Glenn K. Otis et al., *Assessment of Mission Kill Concept, Requirements, and Technologies, Final Report* (Defense Research Projects Agency, Arlington, VA, September 1990).
6. *Policy for Non-lethal Weapons,* U.S. Department of Defense Directive 3000.3, signed by John P. White, Deputy Secretary of Defense, 9 July 1996.
7. NATO AGARD Study AAS-43, "Minimizing Collateral Damage During Peace Support Operations," October 1996.

2. *Are We the World's Police Force?*

1. Sean D. Naylor, "Somalia Revisited: Is the Army Using Any of the Lessons It Learned?" *Army Times,* 7 October 1996.
2. Peter Arnett, A television interview with Osama Bin Laden shown on *Impact,* CNN, 10 August 1997.
3. George J. Church, "Anatomy of a Disaster," *Time,* 18 October 1993.
4. Sean D. Naylor, "Washington Held Back Gunships: Ground Commanders in Somalia Bickered with Policy-Makers in Debate Over Arms," *Army Times,* 16 October 1995.

5. Patrick Pexton, "'Mission Creep' Dubbed Killer of Public Support," *Army Times,* 20 November 1995.
6. Chris Lawson, "Words from a Rising Star," *Defense News,* April 1995.
7. "François Duvalier," *Compton's Living Encyclopedia,* Compton's Learning Company, 1996, Online, America Online (7 July 1997).
8. Gustavo Antonini et al., "History of Haiti," *Compton's Living Encyclopedia,* Compton's Learning Company, 1996, Online, America Online (7 July 1997).
9. Commandant James O. ("Oke") Shannon, U.S. Navy (retired). Oke Shannon and I served in the same group at Los Alamos National Laboratory. In Vietnam he had served with the Navy's riverine patrols, thus he had seen Third World poverty. From 1995 to 1997 Oke served as an adviser to General Sheehan and traveled to Haiti. In several private conversations we discussed the abject poverty found there.
10. Fred Reed, "Chaos is Daily Fare in Haiti," *Army Times,* 9 January 1995.
11. Rick Bragg, "Soldiers Not Prepared for Conditions in Haiti," *The New York Times Magazine,* 6 November 1994.

3. Emerging Threats

1. The force structure for this scenario is taken from National Defense University's "1997 Strategic Assessment: Flashpoints and Force Structure," Chapter Seven, "Persian Gulf."
2. *Washington Post,* 25 January 1997.
3. Douglas Waller and Tom Morganthau, "The Hunt Begins," *Newsweek,* 8 March 1993.
4. *The American Banker,* 1 March 1993.
5. *Newsweek,* "The Week," 1 May 1995.
6. *Newsweek,* 28 August 1995.
7. Pharis Williams, an explosives expert at New Mexico Tech, in private conversation with the author. He notes that the original estimate was 1,500 pounds. However, after tests Williams and others conducted on the New Mexico Tech ranges, the figure was more than tripled.
8. Tom Morganthau et al., "The View from the Far Right," *Newsweek,* 1 May 1995.
9. Richard Pyle, "Suspects in Subway Plot Tied to Hamas," Associated Press, 2 August 1997.
10. Steven Starrier et al., "A Cloud of Terror—And Suspicion," *Newsweek,* 3 April 1995.
11. *The Wall Street Journal,* 21 March 1995.
12. *The New York Times,* 29 March 1995.
13. Robert Kupperman and David Smith, "Coping with Biological Terrorism," Los Alamos National Laboratory White Paper, 22 August 1991.
14. Phil Williams, "Transnational Criminal Organizations: Strategic Alliances," *The Washington Quarterly,* Winter 1995.
15. Robert H. Kupperman, "Executive Summary," *Global Organized Crime: The New Evil Empire,* Center for Strategic and International Security, Report 1994.
16. Douglas Farah, "Russian Mob, Drug Cartels Joining Forces," Washington Post Foreign Service, 29 September 1997.
17. Reuters Limited, "Russian Scientist Backs Claim of 'Suitcase Nukes,'" CNN Interactive, 2 October 1997.

18. James Woolsey, "Global Organized Crime: Threats to U.S. and International Security," *Global Organized Crime: The New Evil Empire,* Center for Strategic and International Security, Report 1994.
19. Robert H. Kupperman, "Trends in Global Organized Crime: Additional Observations," *Global Organized Crime: The New Evil Empire,* Center for Strategic and International Security, Report 1994.
20. Senator John Kerry, *The New War: The Web of Crime that Threatens America's Security* (New York: Simon & Schuster, 1997).
21. Scott Sandlin, "Navy Unit, APD Sued for Drug Raid Death: Widow's Complaint Cites Rights Violations," *Albuquerque Journal,* 7 October 1992.
22. Jessica Matthews, "Power Shift," *Foreign Affairs,* vol. 76, no. 1, January/February 1997.
23. Ibid.
24. Ibid.
25. Claude L. Inis, Jr., "The United States and Changing Approaches to National Security and World Order," *Naval War College Review,* vol. XLVIII, no. 3, Summer 1995.
26. Hans Binnendijk and Patrick Clawson, "New Strategic Priorities," *The Washington Quarterly,* vol. 18:2, Spring 1995.
27. Richard H. Shultz, Jr., and J. Marlow Schmauder, "Emerging Regional Conflicts and U.S. Interests: Challenges and Responses in the 1990s," *Studies in Conflict and Terrorism,* vol. 17, 1994.
28. Robert D. Kaplan, "The Coming Anarchy," *The Atlantic Monthly,* February 1994.
29. David Rohde, "Ted Turner Plans $1 Billion Gift for U.N. Agencies," *The New York Times,* 19 September 1997.
30. Associated Press, "Two North Korean Defectors Describe Kim as Firmly in Power," Washington, D.C., 27 September 1997.

4. *Law Enforcement*

1. The choice of intentionally inducing deprivation is highly questionable. Since Koresh was believed to be mentally unstable, depriving him of sleep, especially since authorities knew he was recovering from gunshot wounds, does not make sense. For the rationale the negotiators were providing him, it would be desirable to have an individual who could think clearly. Also there were still twenty-five children inside.
2. Only one report on children exposed to CS was available. That exposure lasted two hours, not six, as in the Waco case. Other problems with CS, specifically that the carrier, methylene chloride, is toxic to the central nervous system and lowers the flash point of other inflammable substances, were apparently not presented.
3. Although child-abuse questions had been raised in the past, the local sheriff, and others who investigated the claims, found them to be groundless. It is believed by many people that the FBI agents in Washington used false accusations to gain the sympathy of Attorney General Reno. In reality, the HRT was getting tired and wanted to leave, but that was not a sufficient reason to justify the final raid.
4. Gerry Spence, *From Freedom to Slavery* (New York: St. Martin's Press, 1993 and 1995). Gerry Spence, an internationally known lawyer, defended Randy Weaver. This book recounts the facts in this case in graphic detail.
5. *The Wall Street Journal,* 10 January 1995.

6. David G. Boyd, Presentation at the Non-Lethal Defense Conference II, 6 March 1996, McLean, Virginia.
7. Timelines are from the 1986 report of the Mayoral MOVE Commission.
8. *Philadelphia Inquirer* on Philadelphia Online.

5. *Electromagnetic Weapons*

1. For those of us who are proponents of non-lethal weapons, this announcement was a high-risk strategy and, frankly, invoked fear. These were new weapons systems, and training was, at best, modest. If the weapons were employed and failed in this early test, all our efforts could have been trashed. Fortunately, that did not happen.
2. STINGRAY was a medium-power laser developed to craze optical sensors on enemy armored vehicles. Its purpose was to allow outnumbered U.S. and NATO armed forces to prevent Soviet tanks temporarily from acquiring our tanks, while we sequentially killed accompanying tanks. Given the power of the laser, eyeballs behind an optical sensor would most likely be blinded. These prototypes were in engineering development at the time of Desert Storm but sent over anyway to counter the optics on the large number of Iraqi tanks.
3. N. Cook, "Chinese Laser Blinder Weapon for Export," *Jane's Defence Weekly,* 27 May 1995.
4. John Barry and Charles Rogers, "Sea of Lies," *Newsweek,* 13 July 1992.
5. Tactical options were available, however. Two U.S. Navy F-14 Tomcats were in the area and could have responded. They would have been able to observe the aircraft and make positive ID. Captain Rogers ordered them to stay out of the area. As a result of his actions, the captain brought terrorism home with him. The Iranians retaliated directly against him. A bomb was placed in a van driven by his wife. She survived the attack that occurred in San Diego, California.
6. AGARD Study AAS-40, "Non-Lethal Means for Diverting or Forcing Non-cooperative Aircraft to Land," completed July 1995. I was a U.S. delegate to this study group.
7. "Victims of the Shoot Down," *Air Force Times,* 3 June 1995.
8. "Pilots Use Faulty IFF Equipment," *Air Force Times,* 4 October 1995.
9. Captain Jim Wang, of the AWACS crew, was court-martialed and acquitted. Captain Erick Wickson, one fighter pilot, was granted immunity for testimony. Charges were filed against Lieutenant Colonel Randy May, the other fighter pilot, but were later dropped. Captain Cleon Bass, USAF, father of Spec. Cornelius Bass, one of the victims, believing the Air Force was responsible for his son's death, resigned his commission after nineteen years of service.
10. This technology was under contract for the Defense Research Project Agency (DARPA) and U.S. Army Armaments Research and Development Engineering Command (ARDEC).
11. Robert Holzer and Neil Munro, "Microwave Weapon Stuns Iraqis," *Defense News* 13–19 April 1992. This article claimed that the Iraqi air-defense systems were shut down by an HPM warhead on a U.S. Navy Tomahawk missile. *Defense News* prides itself on staying on top of advances in technology. In this case, *Defense News* actually got ahead of the field.
12. David A. Fulghum, "ALCMS Given Nonlethal Role," *Aviation Week & Space Technology,* 22 February 1993.

13. David A. Fulghum, "EMP Weapons Lead Race for Non-Lethal Technology," *Aviation Week & Space Technology,* 24 May 1995.
14. Carlo Kopp, "The E-Bomb—A Weapon of Electrical Mass Destruction," Monash University, Clayton, Australia, posted on the World Wide Web at *infowar.com.* This is a very complete reference, including design criteria for system development.
15. Gregg Meyer, "Ode to the Taser Gun," *Los Angeles Daily Journal,* 22 April 1991.
16. "Survey of Limited Effect Weapons, Munitions, and Devices," developed by Battelle under an Advanced Research Projects Agency contract for U.S. Special Operations Command, 17 July 1995. This publication provides details on hundreds of non-lethal weapons that are available off-the-shelf, or with minor additional development.
17. George Hathaway, private communications over the past 15 years.
18. David Jones, "Israel's Secret Weapon?" *Weekend* magazine, Toronto, 17 December 1977.

6. *Chemical Options*

1. Mark Nollinger, "Surrender or We'll Slime You," *Wired,* February 1995, was the article with this specific title. Others have been written with similar names. Sticky foam has been featured in most such articles.
2. Peter B. Rand, "Foams for Barriers and Non-Lethal Weapons" *Proceedings of the SPIE Security Systems and Non-Lethal Technologies for Law Enforcement Conference,* Boston, 21 November 1996. The author chaired this conference on non-lethal weapons.
3. Steven H. Scott, "Sticky Foam as a Less-Than-Lethal Technology," *Proceedings of the SPIE Security Systems and Non-Lethal Technologies for Law Enforcement Conference,* Boston, 21 November 1996.
4. Peter Rand, in private conversation with the author, 21 November 1996.
5. Charlie A. Beckwith, *Delta Force* (New York: Dell, 1983).
6. "Benign Intervention Technologies Non-Lethal Weapons," Fraunhofer-Institute for Technology Trend Analysis conducted a workshop for the German Ministry of Defense and published a comprehensive report, December 1995.
7. General Glenn K. Otis et al., "Assessment of Mission Kill Concept, Requirements, and Technologies," Defense Advanced Research Projects Agency Report SPC 1361, September 1990.
8. Steve Scott in private conversation, 5 August 1997.
9. The chemical names for these substances are CN—alpha-chloroacetophenone; CS—ortho-chlorobenzylidene malononitrile; CM—diphenylamine arsenic chloride; and OC—oleoresin capsicum.
10. This had relevance in the Waco case, in which very high concentrations of CS were used in an enclosed area on young children.
11. Scott, private conversation.
12. Thommy D. Goolsby, Sandia National Laboratory, "Aqueous Foam as a Less-Than-Lethal Technology for Prison Applications," *Proceedings of the SPIE Security Systems and the Non-Lethal Technologies for Law Enforcement Conference,* Boston, 21 November 1996.
13. Fraunhofer-Institute for Technology, trend analysis.
14. In the most infamous case of its kind, Dr. Frank Olson, a U.S. Army employee,

was given LSD without his knowledge. Some time after that, he exited the twelfth-floor window of a New York City hotel.

15. Michael Crichton, *The Lost World* (New York: Alfred A. Knopf, 1995). Crichton incorporated several non-lethal weapons into this book.

7. *Physical Restraints*

1. This example is not hypothetical. It really occurred and was filmed in Europe as a training measure to demonstrate the effectiveness of simple net technology.
2. *Escape from L.A.,* Paramount Studios, released 1996, starring Kurt Russell as Plisskin.
3. Netgun is a trademark of Capture Systems Inc., who holds the patents on this device.
4. Foster-Miller has created a family of net alternatives. They may be contacted at 350 Second Avenue, Waltham, Massachusetts 02154-1196.
5. From private discussion with Sir Ronald Oxbrough, then Chief Scientist for UK MOD.
6. DARPA-SOCOM book, developed by Alliant Techsystems, page 301.
7. Ibid.
8. NATO AGARD Study AAS-40, "Non-Lethal Means for Diverting or Forcing Non-cooperative Aircraft to Land," completed July 1995.
9. Arnis Mangolds, *The Net Alternative,* Foster-Miller, Inc., September 1994
10. Arnis Mangolds, Foster-Miller, in private conversation, August 1997.
11. Spyder-gun, conceptualized by Craig Taylor and others in the Chemical Science and Technology Division at Los Alamos National Laboratory.
12. Donna Marts is the technical developer of several non-lethal technologies. She currently works at Idaho National Laboratory.

8. *Low Kinetic Impact*

1. This is a fictitious incident. The signal is taken from the Palm Beach County Sheriff's Department, where my son works. Other departments have different signals for domestic cases.
2. Unfortuantely, the 911 emergency number is used excessively for nonemergencies. Delays of several minutes for critical calls are not unusual.
3. "Girl Calls in Her Own Death Plot, Carries It Out," Associated Press, 23 November 1996.
4. "The Use of Plastic Bullets in Northern Ireland," *Congressional Briefing Paper,* April 1993.
5. David E. Steele, "Less than Lethal," *Law and Order,* January 1995.
6. "Non-lethal Ordinance Family of Products," TAAS-Israel Industries Ltd. Contains complete descriptions of each of the weapons made by them.
7. After the American Civil War an eighty-joule rule (fifty-eight foot-pounds) was established. It defined the amount of energy necessary for a penetrating projectile to remove a soldier from the battlefield.
8. "Survey of Limited Effects Weapons, Munitions, and Devices," US Special Operations Command and ARPA, 2nd edition, 29 December 1995.
9. Jamie H. Cuardos, "Definition of Lethality Thresholds for KE Less-Lethal Pro-

jectile," *SPIE Proceedings, Security Systems and Nonlethal Technologies for Law Enforcement,* 19–21 November 1996.

10. Charles Byers. Information is taken from Accuracy Systems Ordnance Corporation literature and private communications.
11. Baton Ball is manufactured by ISTEC Services in the United Kingdom.
12. STINGBALL is also manufactured by Accuracy Systems Ordnance Company of Phoenix, Arizona.
13. The ring airfoil grenade was developed by U.S. Army Chemical Research and Development Command at Aberdeen Proving Grounds, Maryland.
14. Examples of emotional rhetoric can be found posted in *An Phoblacht,* the newspaper of the Irish Republican Army. They also have a Web site that addresses the use of rubber and plastic bullets.
15. David Viano et al., "Mechanism of Fatal Chest Injuries by Baseball Impact: Development of an Experimental Model," *Clinical Journal of Sports Medicine,* 2: 1992.
16. David K. DuBay et al., "Biomechanics and Injury Risk Assessment of Less Lethal Munitions: Analysis of the Defense Technology #23BR Bean Bag," internally published report by Defense Technology, 1996.

9. Acoustics

1. The details of this battle are set forth in the Book of Joshua, Chapter 20, King James Version of the Holy Bible. This may be the first example of the use of nonlethal technology to increase the lethal consequences.
2. Prior to becoming CINC Southern Command, then-Lieutenant General Thurman had been in charge of recruiting and retention for the entire Army but was known to be out of touch with young soldiers. It was Max Thurman who was credited with coming up with the slogan "Be All You Can Be." He hired J. Walter Thompson, a Madison Avenue advertising firm, to craft his ads. He would sometimes object and think their approach was wrong. In turn, Thompson executives pointed out that the ads were geared to recruiting young people. As they told him, "Retention of lieutenant generals is really not a problem."
3. O. Backteman, J. Kohler, and L. Sjoberg, "Infrasound—Tutorial and Review," Report number TR 7.437.04, Swedish Defence Material Administration, 14 July 1983.
4. Dr. Gonzalo M. Diaz, "Safety in Training and Research," an official statement of the American Institute of Ultrasound in Medicine, March 1993.
5. Vladimir Gavreau, "Infrasound," *Science Journal,* January 1968.
6. H. M. Trent, "The Production of Intense Audio Sounds by an Intermittent Flame," Combined Intelligence Objectives Sub-Committee, undated. Taken from the interrogation of Dr. Wallanschek at the end of World War II.
7. Robert P. Shaw, "Methods and Means for Producing Physical, Chemical, and Physiochemical Effects by Large-Amplitude Sound Waves," U.S. Patent 3,087,840, patented 30 April 1963.
8. Guy Obolensky, in private conversation, June 1997.
9. Infrasound is found at very low frequencies. While there is not total agreement, it is generally accepted to be in the range from 0 to 20 hertz. This is far below the audible range.
10. Executive Summary, SARA document on acoustic weapons, March 1997.

11. This work, which I sponsored, was done under internal LANL funding. Due to lack of support from upper management in the Defense Program Office, the effort was prematurely terminated.
12. Curt Larsson and Bengt Wigbrant, "Acoustics as a Non-Lethal Weapon," German-Sweden joint workshop on non-lethal weapons, 12–14 December 1995.
13. John Dering, of SARA, private conversation about recent tests results.
14. Larsson and Wigbrant.
15. Lieutenant Colonel John D. La Mothe, "Sound as a Means of Altering Behavior," *Controlled Offensive Behavior,* Defense Intelligence Agency, ST-CS-01-169-72, July 1972.
16. SARA internal documents were provided for this manuscript. Many of the specific design characteristics are still considered to be proprietary.
17. Henry Kjellson, Forsvunden Teknik, Copenhagen, Nihil, 1974.
18. Dean Barker and I have had several private conversations about this technique.
19. G. V. Batanov, "Characteristics of Etiology of Immediate Hypersensitivity in Conditions of Exposure to Infrasound," undated.
20. SARA briefing paper on acoustic programs, March 1997, and private discussions with the parties involved.

10. Information Warfare

1. Tom Clancy, *Debt of Honor* (New York: Putnam, 1994).
2. Clancy's patriotic stance is well known throughout the military. As guest speaker at Non-Lethal Defense II conference, he received a standing ovation when he thanked the audience, comprised of military officers and supporting industrialists, "for winning World War Three."
3. M. J. Zuckerman, "U.S. Networks Most Vulnerable of Any Nation," *USA Today,* 19 March 1997.
4. Roger C. Molander, Andrew S. Riddile, and Peter A. Wilson, "Strategic Information Warfare: A New Face of War," *Parameters, US Army War College Quarterly,* August 1996.
5. Winn Schwartau, *Information Warfare: Chaos on the Electronic Superhighway* (New York: Thunder's Mouth Press, 1994).
6. Toffler, Alvin and Heidi, *War and Anti-War: Survival at the Dawn of the 21st Century* (New York: Little, Brown, 1993).
7. "Report of the Defense Science Board Task Force on Information Warfare-Defense (IW-D)," Office of the Undersecretary of Defense for Acquisition & Technology, 8 January 1997.
8. General Ronald R. Fogelman, Air Force Chief of Staff, "Information Operations: The Fifth Dimension," remarks to the Armed Force Communications-Electronics Association, Washington, D.C., 25 April 1995.
9. Martin Libicki, "What Is Information Warfare?" Institute for National Strategic Studies, National Defense University, Washington, D.C., August 1995.
10. Dennis Richburg, "Information Warfare," a briefing presented at Non-Lethal Defense I conference, Johns Hopkins, November 1993.
11. Molander, "Strategic Information Warfare."
12. Michael Vlahos et al., "Policy Forum: Pearl Harbor in Information Warfare?" *The Washington Quarterly,* Center for Strategic and International Studies, vol. 20, no. 2. Spring 1997.

13. "Report of the Defense Science Board Task Force on Information Warfare-Defense (IW-D)," Office of the Undersecretary of Defense for Acquisition & Technology, 8 January 1997.
14. DOD Joint Publication, *Information Warfare,* Office of the Joint Chiefs of Staff, 1997.
15. Ibid.
16. Molander, "Strategic Information Warfare."
17. Libicki, "What Is Information Warfare?"
18. Robert Kupperman and Frank Cilluffo, "Between War and Peace: Deterrence and Leverage," *The Brown Journal of World Affairs,* vol. IV, issue 1, Winter/Spring 1997.
19. Cliff Stoll, *The Cuckoo's Egg: Tracking a Spy Through the Maze of Computer Espionage,* (New York: Pocket Books, 1990).
20. Dennis Fiery, *Secrets of a Super Hacker by the Knightmare* (Port Townsend, WA: Loompanics Unlimited, 1994).
21. Dan Scholes, "Kevin Mitnick: The Most Notorious Hacker" (Scholedm@webster2.websteruniv.edu).
22. John F. Harris, "Panel Urges Federal Government to Step Up Efforts Against Computer Terrorism," *The Washington Post,* 21 October 1997.
23. Senator John Kerry, *The New War: The Web of Crime that Threatens America's Security* (New York: Simon & Schuster, 1997).
24. Molander, "Strategic Information Warfare."
25. "War in the Information Age," published by the Institute for Foreign Policy Analysis, Washington, D.C., April 1996.
26. Dr. Eric Davis, in private conversations after his return from spending a year in Korea teaching physics and computer science.
27. The information about IW attacks comes mainly from the people I worked with on these projects at Los Alamos. Unfortunately, that group is now defunct. Of necessity, these scientists choose to remain nameless. You are better off because they were there.
28. Commander Richard P. O'Neill, "Toward a Methodology for Management Perception," Naval War College, June 1989.
29. John L. Petersen, "Info Wars," U.S. Navy *Proceedings,* May 1993.
30. *Joint Doctrine for Military Deception, Joint Pub 3-85,* Office of the Joint Chiefs of Staff, 6 June 1994.
31. Colonel Richard Szafranski, "A Theory of Information Warfare: Preparing for 2020," *Air Chronicles,* USAF Air University (editor@max1.au.af.mil).

11. Biology

1. Gregory Benford, "The Biological Century," an invited presentation to the annual meeting of the Society for Scientific Exploration, Las Vegas, Nevada, 7 June 1997.
2. "Overview: Biological Warfare Issues," *Battlefield of the Future,* U.S. Air Force Air University.
3. Lieutenant Colonel Terry N. Mayer, "The Biological Weapon: A Poor Nation's Weapon of Mass Destruction," *Battlefield of the Future,* U.S. Air Force Air University.
4. "Overview: Biological Weapons Issues."
5. Dr. Dane Jones, "Biological Warfare and the Implications of Biotechnology," California Polytechnic Institute, Course Chem 450—Chemical Warfare (1997).

6. "U.S. Funds Research at Once-Secret Russian Germ Lab," *Newsday,* 10 August 1997.
7. Colm Kelleher, a biochemist and colleague at the National Institute for Discovery Science, in private conversation.
8. Jones, "Biological Warfare and the Implications of Biotechnology."
9. Ibid.
10. Christopher Green, M.D., Ph.D., private communication. Kit Green is renowned for his expertise in toxicology, specifically as related to chemical and biological warfare. The evidence for this BW attack includes the names of scientists involved, from sources with direct access to the research laboratories, flight characteristics of the aircraft that delivered the BW agent, and confirmation from communications intercepts from three countries.
11. J. P. Roberts, M.D., "Ebola Zaire 1976 & 1995," on Internet at virology@net, 14 May 1995.
12. C. J. Peters, "Viral and Bacterial Outbreak Time Line," Centers for Disease Control, Atlanta.
13. Anita Manning, cover story, *USA Today,* 13 May 1997.
14. Associated Press, "Drug Resistant TB Poses Global Threat," 23 October 1997.
15. During the Vietnam War, venereal diseases were rampant in Southeast Asia. It was not unusual for soldiers to have multiple cases of VD in a single year. One favorite method to prevent, or limit severity, was to take continuous doses of antibiotics, thus called "no-sweat pills." This had the effect of introducing new, resistant strains. In many ways, technically developed countries have continued that process. In the future, proper use of antibiotics will be a social responsibility.
16. Tara Meyer, "U.S. Hospitals on Alert After Discovery of Drug-Resistant Germ," Associated Press, 22 August 1997.
17. Lieutenant Colonel Robert P. Kadlec, "Biological Warfare for Waging Economic Warfare," *Battlefield of the Future,* Air University.
18. "Cuba Accuses U.S. of Insect Warfare," Associated Press, 25 August 1997.
19. Office of Technology Assessment, "Proliferation of Weapons of Mass Destruction," Washington, D.C. Government Printing Office, 1993.
20. Lieutenant Colonel Terry N. Mayer, "The Biological Weapon: A Poor Nation's Weapon of Mass Destruction," *Battlefield of the Future,* U.S. Air Force Air University. In this chapter Lieutenant Colonel Mayer details the history of biological warfare and states the case for the threat from biological weapons. He notes the efforts that are underway to reduce that threat.
21. Robert P. Kadlec, "Twenty-first Century Germ Warfare," *Battlefield of the Future,* U.S. Air Force Air University. In the following chapter, Lieutenant Colonel Kadlec develops the future of biological warfare and compares the consequences with chemical weapons. He outlines the intelligence and medical defensive measures and addresses the complex policy issues. Kadleck concludes that biological warfare cannot be narrowly focused on whether or not people are killed by such weapons, but must include the potential for devastating economic losses and the resultant political instability.
22. Pat Unkefer is a former deputy group leader at Los Alamos National Laboratory. We held many conversations about what it would take to develop new organisms.
23. Captain James R. Campbell, Ph.D., U.S. Navy, "Defense Against Biodegradation of Military Materiel," *Proceedings of Non-Lethal Defense III,* 26 February 1998.

24. Raul Cano, "Revival and Identification of Bacterial Spores in 25- to 40-Million-Year-Old Dominican Amber," *Science,* vol. 268, 19 May 1995.
25. Richard Danzig and Pamela B. Berkowski, editorial, *Journal of the American Medical Association,* August 1997.
26. Anita Manning, "Expect Germ Terrorism, Experts Warn," *USA Today,* 6 August 1997.
27. "Pentagon Prepares to Fight Off a Chemical Hit," *Philadelphia Inquirer,* 7 August 1997.

12. *Peace Support Operations*

1. "Minimizing Collateral Damage During Peace Support Operations," *AGARD Advisory Report 347,* May 1997.
2. "Operations Other Than War," Chapter 13 of "Military Operations," *U.S. Army Field Manual FM100-5,* June 1993.
3. "Peace Operations," *U.S. Army Field Manual FM100-23,* June 1993.
4. Associated Press, "Military's Evacuation Role Expanding," *Las Vegas Review-Journal and Las Vegas Sun,* 1 June 1997.
5. The World Court is the popular name for the UN International Court of Justice, established in 1945. It is this body of judges that indicts war criminals.
6. Associated Press, "Arrest of War Crimes Suspect Moves Serbs to Close Ranks," 13 July 1997.
7. Misha Savic, "NATO-led Troops Seize Several Thousand Weapons from Police Loyal to the Rival of Bosnian Serb President Biljana Plavsic," Associated Press, 21 August 1997.
8. Colonel Johan Hederstedt et al., *Nordic UN Tactical Manual,* vol. 1, 1992.
9. *The Army Field Manual, Volume 5, Operations Other Than War, Part 2, Wider Peacekeeping,* HMSO, United Kingdom, 1995.
10. Steven Erlanger, "How Bosnia Set Stage for Albright-Cohen Conflict," *The New York Times,* 12 June 1997.
11. *Army Vision 2010: The Geostrategic Environment and Its Implication for Land Forces,* U.S. Army 1997.

13. *Technological Sanctions*

1. Kimberly A. Elliott and Gary Hufbauer, "'New' Approaches to Economic Sanctions," in U.S. Intervention Policy for the Post–Cold War World: New Challenges and New Responses, the Eighty-fifth American Assembly, Harriman, New York, April 1994.
2. Sanctions constituted the point of departure for our discussion of the contribution of non-lethal weapons at that meeting.
3. Capabilities without the will to use them are useless. Saddam Hussein based his resistance to U.S. and UN forces on his misguided notion that the United States in particular did not have the will to endure casualties.

14. *Strategic Paralysis*

1. There are many variations to the Western spelling of Mohmmar Qadaffi. Even news agencies have not settled on a single version.

2. Robert F. Dorr, "El Dorado Canyon Still Resonates," *Air Force Times,* 22 April 1996.
3. "The Personal History of Mohmmar Qadaffi," emernet.emergency.com/qaddifi [*sic*]
4. "Libya: International Terrorism and Support for Insurgent Groups," Director of Central Intelligence, cs@field(DOCID+ly0159)at lcweb2.loc.gov.
5. "Overview of State-Sponsored Terrorism, 1995," *US Department of State, Patterns of Global Terrorism, 1995,* April 1996.
6. "Huge Chemical Arms Plant Near Completion in Libya, U.S. Says," *The New York Times News Service,* 24 February 1996.
7. President William J. Clinton, "Clinton Letter to Congress on Libya National Emergency," 2 January 1997. This policy was initiated by President Reagan by Executive Order 12543, dated 7 January 1986, and has been continued each year since.
8. Secretary of Defense William Cohen, quoted in "Iran Expanding Arsenal, Cohen Warns," Associated Press, 18 June 1997.
9. Col. John A. Warden III, "Strategic Warfare—The Enemy as a System," Maxwell Air Force Base, Air University Press, 1993.
10. Jason B. Barlow, "Strategic Paralysis: An Air Power Strategy for the Present," *Airpower Journal,* Winter 1993.
11. Maj. Jonathan W. Klaaren et al., "Nonlethal Technology and Airpower: A Winning Combination," an Air Command and Staff College research paper, June 1994.
12. *The CIA World Factbook,* 1995. The section on Libya provided most of the statistics about the country. It should be noted that the reported length of paved roads in Libya varies, even within the same documents. It is only important to note that the road network is very limited.
13. David A. Fulghum, "Air Force May Delay JPATS, TSSAM (Carbon-Fiber Weapon)," *Aviation Week and Space Technology,* 19 September 1994.
14. "Development of the Power Supply in Libya," Lahmeyer International GmbH, Frankfurt.
15. John Barry and Evan Thomas, "A What-If Problem," *Newsweek,* 23 June 1997.
16. Jack N. Summe, "PSYOP Support to Operation Desert Storm," *Special Warfare,* October 1992.

15. Hostage/Barricade Situations

1. John L. Plaster, *SOG: The Secret Wars of America's Commandos in Vietnam* (New York: Simon & Schuster, 1997).
2. Major General John Singlaub, private communications, 23 September 1997.
3. Peter Arnett, "The Secret War," *Impact,* CNN, first aired 14 September 1997.
4. Just to be on the safe side, I also called retired General Carl Stiner, the second commander in chief of U.S. Special Operations Command. He confirmed that even today such a technology does not exist.
5. Multiple sources acknowledge that sensors were placed in the milk cartons that were sent into Mount Carmel. These allowed the FBI to eavesdrop on some of the internal conversations.
6. A shot through the left eye would incapacitate that side of the brain. The left side of the brain controls muscle movement on the right side of the body, thus preventing him from getting a shot off.

16. *Limitations*

1. These were private conversations that took place in Los Alamos National Laboratory. Major General Powell was consulting with the Defense Initiative Office, where I worked. While I initially thought he was attacking me and the concepts, his criticisms proved invaluable. I later learned that he was instrumental in arranging for me to brief several senior Air Force officers.
2. By this statement, I do not infer that we should not have dropped the atomic bombs on Japan. As part of an assignment at Los Alamos, I had the opportunity to study the original documents leading up to the decision. All of the pros and cons of a demonstration or a blockade were thoroughly discussed. The mission assigned to the military was to end the war on our terms *at the earliest possible time.* Having fought for five years (our allies even longer), and facing invasions many times more difficult than Operation Overlord at Normandy, with projected casualty figures unacceptable, it was, in my opinion, the right decision. There has been far too much restructuring of history by those who did not live it.
3. Russell Watson et al., "It's Our Fight Now," *Newsweek,* 14 December 1992.
4. Rogers' Rangers were the first American troops to use guerrilla tactics, even before the Revolutionary War. Today's Rangers trace their lineage back to them. John S. Mosby, aka The Gray Ghost, was a Rebel leader who conducted successful hit-and-run operations behind the Northern lines in Virginia and Maryland during the Civil War. Merrill's Marauders ran similar operations against the Japanese troops in Burma during World War II. SOG is an acronym for "Studies and Observations Group." The most highly decorated American unit in the Vietnam War, they operated behind the lines in Laos, Cambodia, and North Vietnam, collecting vital information and disrupting supply routes. In every case listed, the unconventional operations gained such attention as to become part of U.S. military legend.

17. *Strategic Implications*

1. Mickey Edwards, "Decision Making in Washington: A Rule of Thumb," *Behind Enemy Lines,* Chicago: Regnery Gateway, 1983, and lectures 1 September 1993 at Harvard College.
2. President William Clinton, *State of the Union Address,* U.S. Capital, 4 February 1997.
3. Maj. Zafar Abbas was one of approximately one hundred foreign officers in our course that year (1979–1980). In private discussions, many of these officers commented on lack of knowledge of U.S. military leaders about foreign affairs. The year before, one Middle Eastern officer from an oil-rich country offered to fly the class over for a weekend just to show them what it was really like. While he was turned down, it points to the frustration others have with Americans, even well-traveled and educated groups such as field grade military officers.
4. President Clinton has spoken to this issue in his 1998 State of the Union address and his educational reform policies. He has recommended that students be promoted based on demonstrated skills and offered incentives to prevent social promotions.
5. Survey reported on *Morning Edition,* National Public Radio, 7 February 1997.
6. Senator Al Gore, *Earth in the Balance,* (Boston: Houghton Mifflin, 1992). There has

been some controversy over this book. However, Vice President Gore did lay out a number of important issues I have not seen addressed by others. In addition to describing the ecological state of the world, he demonstrates relationships between the environment and politics, economics, religion, and other social forces.

7. Unbeknownst to most Americans, we have had a battalion on a peacekeeping mission in the Sinai Desert since 1973.
8. This statement was made to me by General Meyer in private conversation when we first were discussing non-lethal weapons at his home in 1990.
9. Lieutenant General Anthony Zinni, USMC, in his opening remarks at the Non-Lethal Defense Conference II, 6 March 1996.
10. General John J. Sheehan, USMC, Commander in Chief, U.S. Atlantic Command, and Supreme Allied Commander Atlantic, in his keynote address at the Non-Lethal Defense II conference, 7 March 1996.

18. *In Opposition*

1. Malcolm H. Wiener, Chairman, "Non-Lethal Technologies: Military Force Options and Implications," Council on Foreign Relations, New York, 1995.
2. These words were spoken in my presence at a conference that guaranteed that comments were not attributable. The professor is highly respected in defense policy circles and cannot be taken lightly.
3. Wiener, "Non-Lethal Technologies."
4. When the M-16 rifle began to replace the heavier M-14, many troops were concerned that the light bullet would not have sufficient penetrating power to kill the enemy. It took years before the M-16, with a higher firing rate and capacity for more rounds, was actually accepted by infantry soldiers. Today it is our standard rifle.
5. Besson J. King, "Testing Challenges for Nonlethal Weapons Development," *Proceedings of TECOM Test Technology Symposium '97,* 18 March 1997.
6. Wiener, "Non-Lethal Technologies."
7. Harvey M. Sapolsky, "War Without Killing," a presentation to the National Strategic Forum Workshop on U.S. Domestic and National Security Agendas: "Into the 21st Century," Cantigny, Illinois, 17 September 1992.
8. *Conspiracy Theory* (Warner Bros.) depicts Gibson as a mind-control subject who has been trained as an assassin by renegade CIA scientists.
9. *Nonlethal Weapons,* RDF Productions, TV 4, produced by Cathy Rogers. The statement by Loren Anderson was one of many misrepresentations portrayed in this television program.
10. Lieutenant Sid Heal, Los Angeles Sheriff's Office, private communications. Sid Heal is considered an expert in non-lethal weapons by both law enforcement agencies and the USMC.
11. This information was provided to me by a now retired CIA agent whose identity will not be revealed. However, he is one of very few secret agents who have been captured, and recovered—alive.

19. *Legal Considerations*

1. "The Protocol for the Prohibition of the Use in War of Asphyxiating, Poisonous or Other Gases, and of Biological Methods of Warfare," signed in Geneva, Switzerland, 17 June 1925.

2. Theresa Hitchens, "DOD Nonlethal Effort Fuels Fear of Treaty Violations," *Defense News,* 26 September 1994.
3. J. Goldblat, "Arms Control: A Survey and Appraisal of Multilateral Agreement," SIPRI/Taylor & Francis Ltd., London, 1978.
4. Nick Lewer and Steven Schofield, *Non-Lethal Weapons: A Fatal Attraction?* (London: Zed Books, 1997).
5. Ibid.
6. Ruth L. Sivard, "World Military and Social Expenditures, 1989," Washington, D.C., 1990.
7. "U.S. Funds Research at Once-Secret Russian Germ Labs," *Newsweek,* 10 August 1997.
8. Lieutenant Colonel Margaret-Anne Coppernoll, "The Non-Lethal Weapons Debate," a paper for the Naval Post-Graduate School, April 1997. This paper contains an excellent overview of the legal and moral issues associated with non-lethal weapons.
9. Major Joseph W. Cook et al., "Nonlethal Weapons: Technologies, Legalities, and Potential Policies," *Airpower Journal,* special issue, 1995.
10. Myron Wolbarsht. Private conversation at the Council on Foreign Relations in New York, September 1993.
11. Brigadier General Larry J. Dodgen, "Nonlethal Weapons," *US News and World Report,* 4 August 1997.
12. "Blinding Laser Weapons: Questions and Answers," International Committee of the Red Cross, Geneva, 16 November 1994.
13. Ibid.
14. Robert Burns, "Blinding Laser Arms Draw Fire," Associated Press, 22 May 1995.
15. News release by the Office of the Assistant Secretary of Defense (Public Affairs), Washington, D.C., 1 September 1995.
16. "The Vienna Review Conference: Success on Blinding Laser Weapons but Deadlocked on Landmines," *International Review of the Red Cross,* 1 November 1995.
17. Hays Parks, Special Assistant to the Judge Advocate General of the Army, "Memorandum of Law: *Travaux Preparatoires* and Legal Analysis of Blinding Laser Weapons Protocol," 20 December 1996, and published in *The Army Lawyer,* June 1997.
18. Marc B. Alexander, deputy sheriff, Palm Beach County, Florida, private conversation, March 1997.
19. Convention on the Prohibition of the Development, Production, Stockpiling and Use of Chemical Weapons and on their Destruction, 13 January 1993.
20. Coppernoll in a conversation with U.S. Navy Judge Advocate General, 3 June 1997.
21. Convention at the Prohibition of Military or Other Hostile Use of Environmental Modification Techniques—"ENMOD," 1977.
22. Convention on the Prohibition of the Development, Production, and Stockpiling of Bacteriological (Biological) and Toxin Weapons and on their Destruction, 1972.
23. Jeanie Konstantinou, "Computer Hackers: Invasion of Computer Systems," *Computers and the Law,* 8 December 1995.
24. Ambassador Allen H. Holmes, Assistant Secretary of Defense (SOLIC) in a presentation to Non-Lethal Defense II, Arlington, Virginia, 7 March 1996.
25. Carey Goldberg, "Foes of Land Mines Win Nobel Peace Prize," *The New York Times,* 11 October 1997.

20. Winning

1. President William J. Clinton, *National Security Strategy for a New Century,* The White House, May 1997.
2. This refers to the military "school situation" for an idealized brigade to attack with three battalions in the main attack supported by two battalions in reserve, all focused on securing the high ground or dominating terrain features. The attack is accomplished by firepower, and maneuver and supplies are replenished once the objective is taken.
3. Gerald F. Seib, "Washington Wire," *The Wall Street Journal,* 1 August 1997.
4. "No addresses" refers to future adversaries that are mobile and do not hold to a fixed location. "Us" refers to U.S. citizens protected by laws that differ from how they protect non-U.S. citizens.
5. Colonel John A. Warden, in a dinner speech to the members of the Air & Space Power Symposium at Maxwell Air Force Base, 17 November 1994.

Epilogue

1. Karl Vick and Stephen Buckley, "Base! Base! Terrorism! Terrorism!" Washington Post Foreign Service, 13 August 1998.
2. Michael Grunwald and Vernon Loeb, "U.S. Is Unraveling Bin Laden Network," *Washington Post,* 20 September 1998.
3. "His fatwa was clear: Americans must die, within the next few weeks." ABC News and Starwave Corporation, 10 June 1998.
4. Tim Weiner, "Afghan Camps, Hidden Hills, Stymied Soviet Attacks for Years," *The New York Times,* 24 August 1998.
5. Milton Beardon, interview on National Public Radio, "All Things Considered," 22 August 1998.
6. "Terrorist Facilities Targeted by U.S." The Associated Press, 20 August 1998.
7. Russell Watson and John Barry, "Out Target Was Terror," *Newsweek,* 31 August 1998.
8. The number of missiles fired in each raid is not certain. Reports from news sources vary considerably. The Department of Defense declined to confirm the exact number. Estimates run from a low of sixty and a high of ninety for the training base in Afghanistan and between six and twenty fired at Sudan.
9. "Terrorist Facilities Targeted by U.S." The Associated Press, 20 August 1998.
10. Douglas Davis, "Bin Laden: The war has begun" *The Jerusalem Post,* 25 August 1998.
11. Tim Weiner and James Risen, "Decision to Strike Factory in Sudan Based on Surmise," *The New York Times,* 21 September 1998.
12. Ibid.
13. *Wag the Dog,* New Line Cinema, 1998.
14. David Sanger, "World Finance Meeting Ends With No Grand Strategy But Many Ideas," *The New York Times,* 8 October 1998.
15. General Dennis Reimer, Senate Armed Services Committee Hearings, 29 September 1998.
16. Senator John Glenn, same session.
17. General Charles Krulak, same session.

18. Captain Stephen A. Simpson and Gunnery Sergeant Steven Carlson, "Training for Measured Response," U.S. Naval Institute Proceedings, September 1998.
19. Colonel Joseph Siniscalchi "Non-Lethal Technologies: Implications for Military Strategy" Occasional Paper No. 3, Center for Strategy and Technology, Air War College, March 1998.
20. Cathy Keen, "UF Researcher: Militias are Armed, Dangerous—and Educated," *Science Daily,* 8 May 1998.
21. Lieutenant Sid Heal, Los Angeles Sheriff's Office, private conversations.
22. President William Clinton, "Remarks by the President at the United States Naval Academy Commencement," 22 May 1998.
23. Jeffrey Hunker, "Hearing on the Administration's Program for Critical Infrastructure Protection," House National Security Committe, 11 June 1998.
24. Dr. Ken Alibek, "Terrorist and Intelligence Operations: Potential Impact on the U.S. Economy," Statement before the Joint Economic Council, United States Congress, 20 May 1998.
25. Dr. Ken Alibek, private conversations with author, 8 June 1998.
26. Victor Sheymov, *Tower of Secrets,* (Annapolis: Naval Institute Press, 1993.)
27. Victor Sheymov, "The Low Energy Radio Frequency Weapons Threat to Critical Infrastructure, Statement before the Joint Economic Council," United States Congress, 20 May 1998.
28. Victor Sheymov, private meeting, 9 June 1998.
29. Bradley Graham, "U.S. Studies New Threat: Cyber Attack," *The Washington Post,* 24 May 1998.
30. Carolyn Meinel, "How Hackers Break In and How They are Caught," *Scientific American,* October 1998.
31. William Chesweick, Warwick Ford, and James Gosling, "How Computer Security Works," *Scientific American,* October 1998.
32. Phillip Zimmermann, "Cryptography for the Internet," *Scientific American,* October 1998.
33. Michael Pillsbury, "Revolution in Military Affairs. Part Four," Institute for National Strategic Studies, National Defense University. This document is a compendium of articles written by senior officers of the People's Republic of China and that were translated into English.
34. Harvey Sapolsky and Jeremy Shapiro, "Casualties, Technology, and America's Future Wars," *Parameters,* U.S. Army War College Quarterly, Summer 1996.
35. President William Clinton, "Remarks of the President to the Opening Session of the 53rd United Nations General Assembly," United Nations Headquarters, New York, 21 September 1998.

Appendix A

1. Col. John A. Warden. Now retired, he is still considered one of the U.S. Air Force's brightest strategists. We discussed the notion of parallel war in several private conversations and public meetings.
2. Abraham Maslow, *Toward a Psychology of Being* (New York: D. Van Nostrand Co., 1968).
3. Gidon Gottlieb, "Nations Without States," *Foreign Affairs,* vol. 70, no. 3, May/June 1994.

INDEX